*Hadewijch and Her Sisters*

*SUNY Series,*
*The Body in Culture, History, and Religion*

Howard Eilberg-Schwartz, Editor

*Hadewijch and Her Sisters*

*Other Ways of Loving and Knowing*

John Giles Milhaven

STATE UNIVERSITY OF NEW YORK PRESS

The lines from *Coal* by Audre Lorde, reprinted by permission of W. W. Norton & Company, Inc. Copyright © 1968, 1970, 1976 by Audre Lorde.

The appendix is reprinted from the *Journal of Religious Ethics*, vol. 5, no. 2 (Fall 1977), pp. 157–81, by permission of Religious Ethics, Incorporated.

The essay "A Medieval Lesson on Bodily Knowing," which appeared in *JAAR*, vol. 5, no. 2, is reprinted, in revised version, by permission of the Editor.

Material reprinted from *Hadewijch: The Complete Works*, trans. by Mother Columba Hart, O.S.B. © 1980 by The Missionary Society of St. Paul the Apostle in the State of New York. Used by permission of the Paulist Press.

Production by Ruth Fisher
Marketing by Dana E. Yanulavich

Published by
State University of New York Press, Albany

© 1993   State University of New York

For information, address State University of New York
Press, State University Plaza, Albany, NY 12246

**Library of Congress Cataloging-in-Publication Data**

Milhaven, John Giles.
    Hadewijch and her sisters : other ways of loving and knowing /
John Giles Milhaven.
        p.   cm. — (SUNY series, the body in culture, history, and
religion)
    Includes bibliographical references and index.
    ISBN 0–7914–1541–4  (acid-free). — ISBN 0–7914–1542–2 (pbk. : acid
-free)
    1. Hadewijch, 13th cent.  2. Body, Human.  3. Love.  4. Knowledge,
Theory of.  5. Feminist theory.  I. Title.  II. Series.
B765.H334M55   1993
189'.082—dc20                                                      92–31369
                                                                              CIP

10 9 8 7 6 5 4 3 2 1

To
Dan Maguire, Jim Nelson, Jock Reeder, Bob Springer
for their steady support of my work

# CONTENTS

Introduction                                                                  ix

Abbreviations                                                                 xiii

**Part I**      Hadewijch and the Mutuality of Love                            1

**Part II**     Medieval Women and Bodily Knowing                            73

Appendix    Thomas Aquinas on the Pleasure of Sex
            and the Pleasure of Touch                                        121

Notes                                                                       147

Bibliography                                                                159

Index                                                                       169

It is an open secret. From the beginnings of Western civilization to the present day, certain ways of loving and knowing are extolled by the culture. Other ways of loving and knowing are looked down on by the culture, though they may be admitted to have some necessity or use.

Many times in Western history, people express their experience that some of the scorned or neglected kinds of loving and knowing are precious and important. In the thirteenth century, many women did so, just as in the waning twentieth century, many women are doing so. On the surface at least, it is similar kinds of loving and knowing that, in opposition to the dominant cultural model, these medieval and contemporary women proclaim by word and acts. Hadewijch "of Antwerp," Margaret of Oingt, Elisabeth of Spalbeek, Beatrice of Nazareth seem to agree on certain values with, say, Carol Gilligan (1982), Naomi Goldenberg (1989), Susan Griffin (1978 and 1982), Beverly Harrison (1985) Carter Heyward (1989), Mary Hunt (1991), Audre Lorde (1984), Judith Plaskow (1991), Adrienne Rich, (1976), Rosemary Ruether (1983) and Haunani-Kay Trask (1986).

In this book, I gather evidence for determining what, if anything, they really agree on. Much more comparative study is needed. But one thing is clear. If there are values common to these medieval and contemporary women, the values are not religious. Many of the contemporary women express no religious conviction. Some of the contemporary religious women, like Harrison, Heyward, Hunt, and Ruether, share similar basic values as those who affirm no religious faith. They appeal not only to religious faith but also to common human experience for evidence of these value judgments. The values held in common by a number of medieval and contemporary women, religious or secular, are human values. All the women draw on their human experience to answer a basic, human question: what kind of loving and knowing is supremely worthwhile in itself, a primary, irreplaceable component of human living?

In answering the question, these medieval and contemporary women identify, I believe, a similar kind, or cluster of kinds, of loving and knowing. This kind of loving and knowing has been belittled by their culture but is, in reality, as the women testify from experience, uniquely precious and essential to human life. The ways of loving and knowing so prized by these medieval and contemporary women are all *mutual* and *embodied* knowing and loving.

"Mutual" and "embodied" are terms presently in ethical fashion during the waning twentieth century in the West. Both words are used by many thinkers to disguise as new what are actually old values of the millennial cultural establishment, values that are not particularly mutual or embodied. On the other hand, the thinkers I have read, particularly feminists, who propose as foundational value an "embodied mutuality" that is truly new, challenging to, and in part subversive of traditional thought, do not, it seems to me, go as far as they might in unfolding what this value consists of. I try in this book to advance evidence of what is this "mutual," "embodied" loving and knowing that the women so prize and what are some of its characteristics.

The evidence I bring is principally medieval. Though contemporary women helped direct and focus my attention on experience of profoundly worthwhile embodied mutuality, this book, because of space limitations, presents almost exclusively experiences had and expressed by medieval women. In Part I, we hear Hadewijch, a thirteenth-century mystic and theologian living in the Low Countries, tell of her experience of supreme mutuality of love. In Part II, drawing mainly on social history and art history, we learn of the experience of prized bodily knowing expressed by Hadewijch and other women of her time and place. Even within these perspectives, the procedure of the book is microscopic. It attends to only a small portion of the pertinent experiences of medieval women. I claim only that my selective analysis of minuscule evidence yields promising hypotheses for broader testing.

With contemporary women, I provide only brief correlation: principally by bibliographical reference and by passing quotation. I do not analyze systematically any contemporary writing to show in what respect, if at all, any woman thinker today prizes a similar kind of mutual, embodied knowing and loving as the medieval women did. The present book is a step toward grounding such a comparison. In hope of provoking such a careful comparison, I illustrate occasionally how some medieval and late twentieth-century women coincide strikingly in traits of prized mutuality and embodiment of knowing and loving that they name. I suggest occasionally how the medieval women in traits of their expressed experience may add to the picture drawn by contemporary women of good embodied mutuality. My main suggestion is that medieval and contemporary women converge in raising provoking, promising questions about the riches and reaches of the embodied, mutual human life worth striving for.

It is, for instance, on the basis of erotic experience that certain medieval and contemporary women thinkers argue for embodied mutuality of love as the supreme worth and foundational centrality in human life. Hadewijch considered her experience in some sense "erotic" since it was experience of her love with Divine Love and the word she used for love was that used by the courtly love tradition, *minne*. Hadewijch did not consider her experiential union with Divine Love to be either "sexual" or "bodily," but anyone from the thirteenth or twentieth century reading some of her descriptions of this experience, say, in Vision 7, would find it more similar to sexual experience than to anything else we know.

Consequently, the theology that Hadewijch orchestrated out of this experience harmonizes with and even repeats that of, for example, Carter Heyward, although part of the experience out of which Heyward moves is sexual in the ordinary human sense. It is lesbian loving. Heyward writes:

> To generate friendship—embodied/incarnate mutuality—is the purpose of a sexual theology and ethics, just as it is the heart of a liberating God. (1989, 104)

> I am reflecting upon the erotic as our embodied yearning for mutuality. As such, I am interested not merely in a "theology of sexuality"—exploring the sexual through theological lenses; but rather in probing the Sacred—exploring divine terrain—through sexual experience. (3; similarly 10, 21, and passim.)

> In struggling to come to terms with the pervasiveness of evil in our life together...I have become increasingly interested in probing the character of that which is radically good in our common life: our power in mutual relation as the basis of our hope for the world. This interest has produced this sexual theology. *For, at its core, this book is about nurturing and cultivating the goodness in our lives.* (18)

If we substitute "love" for "sexual," Hadewijch could have written the lines of Heyward just quoted. She would have meant, I submit, much that Heyward means. To trace traits in which the love that Hadewijch experienced and prized above all was a kind of mutuality much like what we call "embodied" and "sexual" is a main objective of the first part of the present book.

At times in the book, I analyze passages of influential Christian thinkers from earliest times and through the Middle Ages, especially texts of Bernard of Clairvaux. I point out how thoroughly self-sufficient and disembodied ideal knowing and loving is for these thinkers. This brings into relief by contrast what in my analysis of medieval women's accounts is new, challenging and suggestive for traditional Western thought. Medieval women seem to have little sense of it, but contemporary feminists are aware that to espouse embodied mutuality as a supreme human good is to take a different road from that of traditional Western thought (e.g., Heyward 1989, 4–7, 16, 94).

I point out in passing how modern thinkers continue faithfully this tradition of Western thought exalting disembodied self-sufficiency as the fontal human good. Early modern thinkers, such as Luther, Kant, and Sartre, do so programmatically, deriving from it immediately a one-way love of all other human beings. Most present-day members of the intellectual establishment, such as Christian ethicists, do it more implicitly. Giving lip service to embodied mutuality or, indeed, sincerely endorsing it as a supreme human good, they often unwittingly give pride of place to self-sufficient, disembodied loving and knowing. But in the present book I offer no documentation or any proof of the assertions of this or the preceding paragraph! I hope only to start and feed a contemporary ethical argument by introducing into the argument voices of medieval women expressing their experience.

In sum, I press in this book historical questions with a philosophical question in mind. My historical analysis aims ultimately at only one thing: how certain medieval women answer a question that grips me and many of my contemporaries: What constitutes the best human living? The question surely grips all human beings, however implicitly. I narrow and sharpen the question: in what respects, if at all, is the best human living mutual and embodied? In what sense is it "mutual" and "embodied"? Often in fruitful, dialogic questioning, we evolve the meaning of our question as we hear answers to it. The generating question of this book evolves as we listen to the medieval women.

In brief, I and any willing readers attempt to start a conversation with some medieval women. As in all conversations, we will inevitably ignore or dismiss or distort much of what we hear. Still, as in all conversations, even those across centuries, we may actually hear something of what the other person is trying to say. The only hope is to ask as clearly as we can and to listen as hard as we can.[1]

## ABBREVIATIONS

CT = "Compendium Theologiae" by Thomas Aquinas

DV = "De Veritate" by Thomas Aquinas

NE = *Nicomachean Ethics* by Aristotle

SBO I = *Sermones super Cantica Canticorum* in *S. Bernardi Opera*, vol. I

SBO II = *Sermones super Cantica Canticorum* in *S. Bernardi Opera*, vol. II

SBO III = *Tractatus et Opuscula* in *S. Bernardi Opera*, vol. III

SCG = *Summa Contra Gentiles* by Thomas Aquinas

ST I = *Summa Theologiae,* I

ST I–II = *Summa Theologiae,* I–II, by Thomas Aquinas

ST II–II = *Summa Theologiae,* II–II, by Thomas Aquinas

ST III = *Summa Theologiae,* III, by Thomas Aquinas

Suppl. = *Supplementum Summae Theologiae*

## PART ONE

Hadewijch and the Mutuality of Love

## 1

Is the history of Western thought, among other things, a story of persistent neglect of certain ways of loving and knowing? Do establishment thinkers consistently reject, ignore, or downgrade these ways of loving and knowing? Are there sometimes a few thinkers who make much of such ways of loving and knowing, even though their voices are soon lost as the main body of Western thinkers marches on? Has this been true for three millennia? Is it true today?

This is a question of historical fact. When and to what extent has this happened in the course of Western thought? In the present book we pursue this historical question by attending to the words and actions of certain medieval women. But our questioning is aimed at and governed by a philosophical question. If certain ways of loving and knowing were and are dismissed programmatically by mainstream Western thinkers, are they valuable ways of loving and knowing? Is their neglect a substantial loss? Is their restoration to fuller use urgent? As I said in the *Introduction*, women thinkers of our time press this historicophilosophical question with precision, power and promise.

In a word, this book scouts historically for an answer to the philosophical question. In Part I we read the writings of a Christian theologian, Hadewijch, who lived in the Low Countries around the middle of the thirteenth century. We attend particularly to ways of *loving* to which Hadewijch gives pride of place. We contrast Hadewijch with other Western Christian thinkers of her time and earlier. Part I, as well as Part II, is a tiny exploration of possible evidence for an affirmative answer to our large historicophilosophical question. May it encourage other studies to follow up, whether to confirm, expand, revise or refute it.

During the twelfth, thirteenth, and fourteenth centuries in Western Europe, human life was changing enormously. People appealed regularly to past precedent, but they regularly—often less consciously—were constructing an unprecedented present. Unprecedented ways of loving and knowing became widespread and powerful. Recall what was going on in connection with universities: songs of courtly love; devotion to the human Christ, Mary, and saints; devotion to the Eucharist; the Magna Charta; codes of chivalry; building of cathedrals; the spread of monasteries and cloisters; new instruments of war; founding of hospitals; a greater variety of imports (including translations from Islamic culture); and a plethora of native biographies and autobiographies. Social historians, such as Caroline Bynum, show new ways of loving and knowing in the data they have gathered on new forms of religious piety, for example, in feasting and fasting, and on new characteristics of domestic life. Art historians, such as Joanna Ziegler, have similarly researched the nature and use of religious art, such as Pietàs in the thirteenth and fourteenth centuries.

In Christian thought of the time, new ways of knowing and loving came into prominence. Much of what was theologically new in the Middle Ages has been traced during the twentieth century by de Ghellinck, Gilson, Chenu, Leclercq, and others. But this advancing mainstream thought did not keep up with lived experience of the time, particularly the experience of devout women (or so I argue through this book). Let us listen to one thinker, an informed, able theologian, paralleling new ways of knowing and loving in popular piety and going ahead of other theologians of her tradition in articulating what was primary in her own experience.

The theologian, Hadewijch, lived in a Beguinage somewhere in the Low Countries during the middle of the thirteenth century.[1] Beguines were devout women largely of noble families, who lived in self-supporting community, and breaking with precedent, chose to live lives of apostolic poverty, contemplation, and care for the sick without taking vows as nuns (Hadewijch 1980b, p. 3). Hadewijch wrote, in the Dutch language of her time, letters, poetry and accounts of her religious, at times mystical, experience. They come to over three hundred pages in Mother Columba Hart's English translation. We know little of Hadewijch or her life. We have no contemporary account of her. Her own writing, though often personal and autobiographical, is vague and reticent on factual detail.

Hadewijch's work was read in some circles and had influence for a century or two. She affected John of Ruusbroec, influential mystical theologian of the fourteenth century, and his disciples. She influenced thus major currents of mystical thought down to the present day. By the sixteenth century we hear no more of her or her works until 1838 when researchers discovered manuscripts of her writings at the Royal Library in Brussels.

Though Jozef Van Mierlo brought out a critical edition of her complete writings between 1924 and 1952, Hadewijch has to the present time received little attention in scholarly literature written in English and not much more in languages other than Dutch. Evelyn Underhill, in her twelfth edition of *Mysticism* (1930), made no mention of Hadewijch. Yet Paul Mommaers has called Hadewijch "the most important exponent of love mysticism and one of the loftiest figures in the Western mystical tradition" (Hadewijch 1980b, xiii). Listen to Letter 9 in its entirety:

> May God make known to you, dear child, who he is, and how he deals with his servants, and especially with his handmaids—and may he submerge you in him. Where the abyss of his wisdom is, [God] will teach you what he is, and with what wondrous sweetness the loved one and the Beloved dwell one in the other, and how they penetrate each other in a way that neither of the two distinguishes himself from the other. But they abide in one another in fruition, mouth in mouth, heart in heart, body in body, and soul in soul, while one sweet divine nature flows through both and they are both one thing through each other, but at the same time remain two different selves—yes, and remain so forever.[2]

In other passages, to be considered below, Hadewijch describes similarly this supreme human experience of the Divine. It is an experience that God grants some of his servants in their earthly life. It anticipates their eternal life with him. The experience, as Hadewijch describes it, resembles the lived experience of other devout persons of her time and the century before. With the help of Bynum and Ziegler, I will point to some of this resemblance in following pages. But Hadewijch's description is also theological. It is theologically articulated and theologically systematic. The theology arises out of and is true to her distinctive experience and that of other women of her time and place. Hadewijch's theology differs correspondingly from all systematic theological accounts of experience of the Divine with which I am familiar before and during Hadewijch's time.

This theologian breaks, therefore, from establishment theology of past and present to express theologically, as well as personally, widespread lived experience of the time. Such expression is typical of medieval theology, for the theologian is here an integral, interactive part of the public community. The theologian influences the community by, for example, teaching clergy and monks and by the clergy's preaching to other monks, nuns, and laity. The influence is mutual, and Bynum and Ziegler point out parallels of theological development and cultural movements.

The quotation given above illustrates the theological articulation that characterizes Hadewijch's pages. Hadewijch does not speak in uniquely narrative terms. In the quoted passage, Hadewijch describes the union of servant and God. She describes the union in general terms, i.e., as applicable to many a "servant." She describes it, too, with general concepts such as fruition, distinguishability, different selves, body, soul, divine nature, and so on. This is true even on other pages where she writes explicitly of her own individual experience of God. Hadewich is a competent, sophisticated theologian who interplays her religious experience and her theology throughout her work. She engages in dialogue with Christian thinkers of the past, echoing, for example, Jerome, Pseudo-Dionysius, Bernard of Clairvaux, and Richard of Saint Victor, while at the same time going beyond and even reversing them.

In the letter quoted above, as throughout her writing, Hadewijch does what all Christian theologians of the West before or during her time did. Their "pagan" forefathers: Platonists, Aristotelians, Stoics, and Neoplatonists did it too. They all work to identify what in their judgment is the completely good human life possible to humans in their earthly existence. It is a union with the divine. It is a specific kind of union.

For Hadewijch, as for Augustine, Pseudo-Dionysius, and other Christian theologians—not to mention pagan, Jewish, and Muslim Platonists and Neoplatonists—before her time, the supreme experience of God that a human being can attain on earth is normative for appraising all human life and activity. The experience is not only satisfying rapture. It is not only overflowing illumination of mind. It is not only exquisite intimacy with God. The supreme experience of God that all these thinkers describe is, for them, the best human knowing and loving possible to humans on earth. All other human knowing and loving, therefore, is of worth only insofar as it resembles or contributes to this supreme experience.

Some of the thinkers so describe this supreme earthly experience that we of the twentieth century might categorize it as mystical. Others so describe it that we might categorize it rather as contemplative. Some thinkers, too, use one or both terms with differing senses. In all cases, this highest possible experience of God is for them a living, however temporary, of the ideal human life. It satisfies all human aspirations. Nothing better is possible for human beings. This experience "surpasses all that one can have from [the Beloved] and all that he himself can accomplish" (Vision 14, 145ff.).[3]

All this is true because, for Hadewijch as for Christian theologians before her, this supreme earthly union with the Beloved is, by God's gift, a momentary anticipation of the union with God that will constitute the beatitude of the soul after death for all eternity. This is why the experience reveals directly, if imperfectly, the supreme good, the final end of humanity, the norm in terms of which all earthly human life is appraised. Anything that a human being does or is on earth is good only insofar as it shares in or moves toward this union which the theologians describe and identify.

In articulating the nature of this union, therefore, the theologians do not only mystical theology but also fundamental ethics. How one of these Western thinkers describes the experience of the supremely possible earthly union with God is a clue to how he or she judges the intrinsic worth of this or that way of knowing and loving. I belabor the point for, as I have already said more than once, this is a foundational angle of my inquiry: What constitutes what is intrinsically good and worthwhile throughout human life? How does Hadewijch answer this question? How does her answer differ from traditional Christian theologians before and during her time? And at times implicitly or in passing: how does her answer differ from dominant Western thought of the late twentieth century, and how does it resemble or add coherently to contemporary feminist thought?

To my knowledge, no Christian theologian before or during Hadewijch's time describes supreme earthly experience with certain traits with which she does. I say this not only of men writing theologically such as Augustine, Bernard of Clairvaux, the Victorines, Thomas Aquinas, Bonaventure, etc. I say this of women, such as Hildegard of Bingen, Beatrice of Nazareth, and Margaret of Oingt who like Hadewijch wrote theologically of and out of their individual experience of the divine. On the other hand, I believe, though I have not carefully verified, that some of Hadewijch's new characterization of these earthly heights of experience is voiced by later Western thinkers, such as Mechtild of Magdeburg, Julian of Norwich, John of the Cross, and Teresa of Avila.

In any case, the burden of Part I is to bring out in her writing some novel traits with which Hadewijch characterizes her supreme experience of God. I contrast Hadewijch's characterization with earlier theologians as Augustine, Pseudo-Dionysius, Bernard of Clairvaux, and Guerric of Igny. My historical searching is exploratory and still very limited. Is Hadewijch's picture of her supreme experience of God as novel as I conclude? I invite other historians of thought to return to further texts and prove or disprove these hypotheses. My main effort in what follows is to record sensitively what Hadewijch tells us that she experienced. I contrast her with certain traditional thinkers, especially Bernard of Clairvaux, mainly in order to bring out more sharply what is this experience she had and describes.

Hadewijch writes about only one thing. She strives to live one thing alone. It is love, and all that love brings with it. Her love, as she believes all human love must be, is to share in Christ's love, God's love. Her love includes, therefore, love for fellow human beings. In this earthly life, love means sacrifice, suffering and service. Hadewijch recounts often her efforts to live this sort of love. She encourages and urges others to do the same (e.g., Letter 6; Vision 1).

Often, however, Hadewijch makes clear that the foregoing traits of love are not those that she most prizes and seeks. She prizes over all else and seeks most passionately with her whole body and soul the final fulfilment of love, i.e., "the fruition of God." She speaks over and over and with passion of "the fruition of God." At times Hadewijch uses the term in a broad sense for the union with God that all who seek and love God can attain on earth, at least to some degree and regularly (e.g., Letter 6, 117ff.,120ff.; Letter 12, 13ff.). At other times, Hadewijch speaks of a "fruition of God" that is vouchsafed on earth only to the few and infrequently to them (e.g., Letter 12, 53ff.; Letter 14, 3ff.).

To be united with God in this supreme fruition of love is the one all-dominant goal for Hadewijch as for all humans. It is what they really want. It is what God wants for them (e.g., Letter 1, 46ff.; Letter 2, 66ff.; Letter 6, 227ff.). It is what Hadewijch with the other blest will have for all eternity: "There I had fruition of him as I shall eternally" (Vision 5, 59ff.; see also 65ff.; Letter 12, 53ff.; Letter 14, 19ff.; Letter 16, 14ff.). She possesses it temporarily in earthly life, though less "amply" (Vision 5, 59ff.; Vision 6, 92ff.; more generally, Letter 12, 13ff.). What I have attributed to Hadewijch in these three paragraphs can be found in many Christian medieval thinkers of Hadewijch's time or earlier.[4]

In some of her writing Hadewijch laments that she has never yet had this fruition (Letter 1, 56 ff.). On other pages, presumably those written later (considered below), she describes her occasional experience of this supreme fruition of God and makes much of its sublime and completely satisfying nature. The fruition consists generally in two stages. First, Hadewijch is taken up "in spirit," "seeing" and "hearing" God. The fruition then changes sharply and climaxes as she comes "out of the spirit" into a new, more complete union with her Beloved, Christ, who is both man and God and thus whose nature is Divine Love:

But then wonder seized me because of all the riches I had seen in him, and through this wonder I came out of the spirit in which I had seen all that I sought; and as in this situation in all this rich enlightenment I recognized my awe-inspiring, my unspeakably sweet Beloved, I fell out of the spirit—from myself and all I had seen in him—and wholly lost, fell upon the breast, the fruition, of his Nature, which is Love. There I remained, engulfed and lost, without any comprehension of other knowledge, or sight, or spiritual understanding, except to be one with him and to have fruition of this union. I remained in it less than half an hour. (Vision 6, 76ff.)

This experience out of the spirit is the supreme union for which, and for which alone, all humans yearn. What humans want is simply their fruition of Divine Love by becoming one with It.[5]

In other passages, too, Hadewijch, as many Western philosophers and theologians before her, describes moments or hours of earthly experience in which the individual unites supremely with God. Visions 5, 7, and 10 to 14 are instances. Her interpretive descriptions of this experience continue and resemble those of prior Christian thinkers. Her descriptions also differ substantially from theirs.

Take, for example, classical descriptions of this kind of experience by two early Christian theologians, two who influenced greatly theologians of the twelfth, thirteenth and fourteenth centuries: Augustine and Pseudo-Dionysius the Areopagite.

Recall in his *Confessions* Augustine's vision as he and his mother stood alone leaning out a window onto a garden at Ostia, "talking alone together sweetly," wondering what would be the future eternal life of the blest.

We were opening the mouths of our hearts toward the streams flowing on high from your fountain, the fountain of life that is with you, so that, sprinkled from it according to our capacity, we could somehow think of so great a reality. Then...raising ourselves with more ardent love to the Selfsame, we proceeded step by step through all bodily things up to the heaven itself whence sun and moon and star shine upon earth. We mounted further thinking of and speaking of and marveling at your works and we came into our minds and went beyond them to reach the region of your unfailing abundance....There...life is the wisdom by which all these things are made that have been and will be. And it is not made,...nor is there in it to have been or to will be, but only to be, for it is eternal....And while we speak and gape at it with longing, we reach it slightly for a complete beat of our hearts....And we sighed...and returned to the noise of our mouths where a word is both begun and ended. And what is like your Word, our Lord, which remains in itself without aging and makes all things new?

Augustine goes on to generalize:

If for a human being [all things changeable] become silent and turn his ear up to him who made them and he alone speak, not through them but through himself, so that we hear his word, not by tongue of flesh nor voice of angel nor thunder of cloud nor puzzle of similitude, but himself, whom we love in these things, himself we hear without them, as we now extend ourselves and with quick thought reach eternal wisdom abiding above us, if this be continued and other visions of far unequal kind be taken away and this one seize and absorb and hide its beholder within its inner joys, so that eternal life would be such as was that moment of intelligence for which we sighed, then is this not to "enter into the joy of the Lord"? (1962, IX, 10, my translation)

In *The Mystical Theology*, Pseudo-Dionysius writes as follows of the supreme union that a human being can attain with God:

The divinest and the highest of the things perceived by the eyes of the body or the mind are but the symbolic language of things subordinate to Him who Himself transcendeth them all. Through these things His incomprehensible presence is shown walking upon those heights of the mind; and then it breaks forth even from the things that are beheld and from those that behold them, and plunges the true initiate into the Darkness of Unknowing wherein he renounces all the apprehensions of his understanding and is enwrapped in that which is wholly intangible and invisible, belonging wholly to Him that is beyond all things and to none else (whether himself or another), and being through the passive stillness of all his reasoning powers united by his highest faculty to Him that is wholly Unknowable, of whom thus by a rejection of all knowledge he possesses a knowledge that exceeds his understanding. (1940, 194)

We are looking for differences between Hadewijch's account of supreme human union with God and accounts by Augustine, Pseudo-Dionysius and other Christian theologians preceding or contemporary with Hadewijch. First, however, note the similarities.

For Hadewijch, as for Western thinkers before her, the supreme union with God is or yields a supreme *knowing* of God (Vision 14, 172ff.; Vision 6, 76ff.; Vision 7, 1ff.; Vision 9, 40ff.; Letter 12, 53ff.; Stanzaic Poem 9, 71–75. For Hadewijch, therefore, and preceding theologians, "perfect knowing" ( *vollen kins*) belongs to the fullness of human life. All human activity and living will be good and worthwhile, to the degree to which it partakes in such knowing. On the other hand, for Hadewijch and earlier theologians, fortunate individuals who achieve to some degree this knowing do so out of love. They know only because they love. Moreover, the knowing leads afterward to greater, stronger, better aimed loving. All human activity and living will be good, worthwhile, to the degree to which it partakes in such knowing and loving.

Hadewijch's fruition of God has, as we saw, two stages. The prior stage, her being taken up "in the spirit," fits the picture (sketched above) drawn by traditional thought. Hadewijch's experience "in the spirit" resembles Augustine's summit, illustrated by the passage from the *Confessions* quoted above. It is the height where divine wisdom abides. There the soul shares voluptuously the wisdom of the divine word. It is an experience of "intelligence" fused with love and joy. For Augustine, the "hearing" is metaphor for knowing by intellect. For Augustine as for all Christians, the divine word is a word of intelligence. The Latin *verbum* continues the Greek *logos*: both terms mean at the same time the thought and the corresponding word. Hart's "spirit" translates Hadewijch's gheeste. The Medieval Dutch word *gheeste*, like the modern Dutch *geest* and German *Geist*, and more strongly than the modern English "spirit," connotes usually intelligence, a noetic dimension. In what Hadewijch describes as her being "taken up in the spirit" she continues the Christian Platonic and Neoplatonic tradition of the supreme heights of intelligence.

In describing this prior stage of fruition, Hadewijch may show originality of thought with respect to her predecessors and contemporaries of the twelfth and the thirteenth centuries, mystics and theologians, women and men. In my study, however, I pass by this prior stage and focus on Hadewijch's final union with God, when she is taken "out of the spirit." This is not further "ascent" but, as in Vision 6 and the passage of Pseudo-Dionysius cited above, a level meeting of God and the individual who has mounted thus far. For Hadewijch, like Pseudo-Dionysius, the uniting on a level is followed by a "falling" onto or "plunge" into the divine. Her "passing away" out of intelligence into an even more intimate union with the divine resembles Neoplatonic ideas, which Augustine and many Augustinians, despite their debt to Plotinus and later Neoplatonists, did not express, but which other Christian theologians did, such as Richard of Saint Victor. This supreme union, though beyond intelligence, is or yields its own unique kind of knowing, as both the Hadewijch and Pseudo-Dionysius passages exemplify.

In describing this second stage of the supreme union with God, this passing "out of the spirit," beyond intelligence, Hadewijch shows a second similarity with earlier thought. This supreme union with God is not first and foremost a knowing of God but, as we saw in Vision 6, a loving of God. It is, above all, a merging of human love with Divine Love. Listen again:

> But then wonder seized me because of all the riches I had seen in him, and through this wonder I came out of the spirit in which I had seen all that I sought; and as in this situation in all this rich enlightenment I recognized my awe-inspiring, my unspeakably sweet Beloved, I fell out of the spirit—from myself and all I had seen in him—and wholly lost, fell upon the breast, the fruition, of his Nature, which is Love. There I remained, engulfed and lost, without any comprehension...except to be one with him and to have fruition of this union. I remained in it less than half an hour. (Vision 6, 76ff.)

In Chapter 4 of *The Divine Names*, Pseudo-Dionysius describes the supreme union with God as a union with divine "love" ( *eros*) (1940, 101–11). So too do twelfth-century thinkers such as Bernard of Clairvaux, who develop the idea more expansively and systematically than Pseudo-Dionysius. Bernard is not so hardily Dionysian as to place this supreme union beyond and "out of intelligence." Bernard locates on the same highest level of human experience both the soul's affective union with God in love and the soul's contemplative union with God in intelligence. But Bernard, like Hadewijch, follows the lead of Pseudo-Dionysius in seeing love as yielding a unique knowing, different from what mere contemplation reveals (e.g., SBO II, XLIX, 4; LXVII, 8; LXIX, 2).[6]

Bernard, more systematically than any theologian before him, affirms, explains and applies this supreme union of human love with Divine Love as the goal of all human endeavor. In so doing he begins the main currents of twelfth- and thirteenth-century theological thought that flow through Hadewijch's writings.[7] This makes her differences from Bernard all the more significant historically and philosophically.

Let us compare more precisely Hadewijch and traditional theologians on the supreme union of the human individual with Love. An "equality," in some sense, with God, as Hadewijch claims for her supreme union with Him, is by her time commonly claimed by theologians; 2 Pet. 1:4, "that you may become sharers of divine nature," gave them warrant. Like Hadewijch, they affirmed at the same time the radical difference of Creator and creature, though other theologians and later historians dispute whether some theologians making such affirmation escape pantheism and thus heresy.

Christian theologians before Hadewijch have affirmed, as she does, that in the supreme union with God, the human person becomes and "is" God. In this respect, Bernard, Hadewijch, and other medieval theologians go beyond Augustine and follow rather Pseudo-Dionysius and other Neoplatonists. Since for Hadewijch, as for Bernard, God is primarily Love, Hadewijch, like Bernard, affirms that in this supreme experience she comes to be Divine Love and God (Vision 1, 138ff.; Vision 3, 1ff.).[8] For Bernard to be God or to be Love is not to lose one's personal identity. It is not to become one with God pantheistically.[9] Hadewijch, too, maintains a personal identity in the final merging. Thus she epitomizes the union as "to be God *with* God" (my emphasis; Vision 7, 1ff.; see Letter 9 above).

To Bernard, for the soul "to be Love" means two things. First, the soul now loves as God loves. It is to love as similarly to God as a human can. It is to will the same as God does. "It is plainly an embrace where to will the same thing and to not will the same thing makes one spirit out of two" (SBO II, LXXXIII, 3; see also elsewhere in 2 and 3; also SBO II, LXIX, 1; LXXI, 7–10; SBO III, *De Diligendo Deo* X, 27–28, 142–43). Hadewijch affirms this, too. God tells her that in this final fruition, "You will be love as I am Love" (Vision 3, 1ff.; similarly, e.g., Vision 14, 145 ff.). She spells it out:

> When the soul is brought to nought and with God's will wills all that he wills, and is engulfed in him, and is brought to nought—then he is exalted above the earth, and then he draws all things to him, and so the soul becomes with him all that he himself is. (Letter 19, 46ff.)

For Bernard, a human being "is Love," secondly, in the sense that he is in the experience unaware of anything else. He is aware no longer of himself, but only of Divine Love (e.g., SBO III, *De Diligendo Deo* X, 27–28, 142–43; SBO II, LXIX, 1; LXXXIII, 3; LXXXV, 13). Similarly, Hadewijch affirms that in supreme union with God she loses all awareness of oneself. "I wholly melted away in him and nothing any longer remained to me of myself" (Vision 7, 94ff.).

The supreme experience Hadewijch has of God is, therefore, in part not new to Christian thought before her time. In part, it *is* new.

Hadewijch depicts the experience with certain traits that neither Pseudo-Dionysius, Bernard, Hildegard of Bingen, nor any other prior or contemporary Christian systematic theologian whom I know did. I say "systematic theologian," for these novel traits Hadewijch asserts of her supreme experience of God are found in accounts of visions and devotion, particularly of women, of Hadewijch's time and the century before. Some of the authors of these accounts of "becoming Divine Love" make theological affirmations, but none that I know of articulate and explain the experience in a coherent theological way. This, Bernard does in his abundant writings and Hadewijch does in her letters, poems and accounts of visions. But Hadewijch, I am about to argue, describes this experience as even Bernard does not.

# 2

What, then, are novel traits which Hadewijch introduces into her theological account of supreme human experience of God? First, for Hadewijch "to love the humanity in order to come to the Divinity" means that in her supreme "fruition" of God she consciously embraces her Beloved as a man as well as God. In her supremely fulfilling fruition she embraces him as one embodied person embraces another. Her embodied embrace of her Beloved is not, as "embrace" ( *amplexus*) is for Bernard, an image or metaphor for something else, something bodiless and spiritual, that she was aware of happening to her. To read her text, say, Vision 7 (quoted below), and Vision 6 and Letter 9, permits only one interpretation: Hadewijch experienced *herself as physically embracing her Beloved.*

Bernard paved a new road for future theologians by endorsing and encouraging a sensible, bodily kind of devotion to the human Christ. But he saw this devotion left behind when Christ comes to the soul for its supreme union with Him. Bernard acknowledged, too, the validity of the kind of experience Hadewijch reports, the bodily visions that pious believers may have of Christ. But he insisted that the supreme union with Christ was not like this. It was in no way bodily. It was far superior (e.g., SBO I, XXXI, XXV, 7; SBO II, LII, 3 and 5; LXXIV, 2, 5–6; LXXXV, 13).

Not so for Hadewijch. Hadewijch does on occasion affirm that she is in her supreme embrace of her Beloved "out of her body" as well as "out of her spirit." But for her, unlike Mechtild of Magdeburg and Margaret of Oingt, the occasion of noting this is rare (Vision 13, 241ff.; Vision 8, 123ff.). In Letter 9, she tells explicitly "with what wondrous sweetness the loved one and the Beloved dwell one in the other, and how they…abide in one another in fruition, mouth in mouth, heart in heart, body in body, and soul in soul." In any case, it is in bodily wise that she experiences her Beloved. A dream, trance, vision, even devout fantasy can be bodily experienced, though the person's actual body is only partly affected. One can be caught up in a fully sensual fantasy of embracing someone although one's actual arms do not reach out and hold. At the moment one is not imaging or comparing; the sensual experience is all one has. Nor does Hadewijch use it subsequently for imagining or comparing her union wth God. It *is* a supreme stage of that union.

For Hadewijch, even the desire for the union is physical: "My heart, soul and senses have not a moment's rest" (Letter 25, 16ff.). Before Vision 7, her heart, veins, and limbs trembled with desire. Her mind was beset so fearfully and so painfully with desirous love that all her separate veins threatened to break and all her separate veins were in travail as she longed to have "full fruition of my Beloved." He came to her in the form of a man, "sweet [*suete*] and beautiful." After he gave Hadewijch the sacrament of the Eucharist, "He came himself to me, and took me entirely in his arms, and pressed me to him; and all my members felt his in full satisfaction, in accordance with the desire of my heart and my humanity" (Vision 7, 64ff.).

Hadewijch's supreme experience of her Beloved is thus, as bodily, of a kind not recognized by traditional theologians for this experience. But "bodiliness" can mean—include or exclude—many things. In Part II, we will try to discern what for Hadewijch, as for other women of her time, this supremely worthwhile bodiliness consists of. In addition, her union with Christ has a *mutuality* unacceptable to theologians of her time and earlier. By force of the New Testament, and indeed the Bible as a whole, Christian theologians differed from Greek and Roman philosophers by attributing considerable mutuality between God and believer. Yet, in the remainder of Part I, I want to trace out how the mutuality which Hadewijch affirms of her supreme union with God goes substantially beyond that affirmed by Christian theologians of or before her time. I want to show in Hadewijch's text what this further mutuality consists in. In Part II I argue that the "excess" of mutuality in this supreme experience is linked to its bodiliness.

Recall the areligious, ethical concern of the present study. As such, it does not concern our inquiry that Hadewijch breaks from theological tradition in describing a full human relationship *with* God. What concerns us is that in so describing she breaks from theological tradition in identifying what characterizes the full loving and knowing, the full living *possible to humans on earth*. For Hadewijch, full human life is preeminently mutual loving and knowing an other.

In this respect, I suggest, and will offer occasional illustration, that Hadewijch affirms something analogous to what women writers—religious, agnostic, or atheist—of the late twentieth century insist on. To the extent that the mutuality which Hadewijch affirms of her fruition of God characterize also fulfilling moments of popular piety of women of her time, Hadewijch also moves with thousands of medieval women, mostly uneducated, in breaking away from the traditional, still dominant Christian ideal that downgrades and subordinates mutuality. She and they by contrast bring out what that dominant ideal was: an ideal of self-sufficiency based on complete dependence.

Some elements of mutuality in Hadewijch's supreme experience are immediately striking. Not only did Hadewijch "wholly melt away" in her Beloved and "nothing any longer remained to me of myself." But she saw her Beloved, too, "completely come to nought and so fade and all at once dissolve that I could no longer recognize or perceive him outside me, and I could no longer distinguish him within me (Vision 7, 64ff., 94ff.). As Hadewijch says in Letter 9 (4ff.), quoted above, "neither of the two distinguishes himself from the other."

That both Hadewijch and the God-Man melt into each other and in their union neither of the two distinguishes himself or herself from the other is uncommon—to my knowledge, unprecedented—in mystical report and theological thought before or during Hadewijch's time. In the Neoplatonic tradition, such assertion made of the soul in its supreme union with the Divine is commonplace. Not commonplace, if made at all, is a corresponding assertion of the Divine, the Good, the One, the finally loved one. In Letter 9 and Vision 7, Hadewijch seems in part to echo Plotinus ( *Enneads VI*, 34), but Plotinus does not say that in union with the soul the divine beloved comes to nought, fades and dissolves. Plotinus does not say that the divine beloved does not distinguish itself from the soul. Nor does Bernard of Clairvaux say these things in his otherwise similar statements of *De Diligendo Deo*, X, 27–28.

In saying what she says, Hadewijch is not pantheistic. Though in their ecstatic awareness, "neither of the two distinguishes himself from the other," Hadewijch maintains that her Beloved and she remain their distinct selves. In the passage from Letter 9 cited below, Hadewijch affirms explicitly that in the most complete union with the Beloved, though they become indistinguishable, he and she remain their distinct selves. Yet in the texts quoted, it is not only of his human form and person that Hadewijch says that her Beloved becomes as indistinguishable and undistinguishing as she. In Letter 9 and Vision 7, as elsewhere (e.g., Vision 6, 76ff.), she speaks explicitly of uniting with and experiencing her Beloved as God as well as Man.

To describe the Divine and Hadewijch as melting into each other so that neither can distinguish the other suggests a temporary equalizing, a relative equalizing of the two persons in experience, novel, I suggest, in Christian mystical thought up through her time. In the relative equality in Bernard's thought noted above, the human person in becoming the Divine person loses all distinct awareness of his or her human self. In the equality Hadewijch affirms, the human person in becoming the Divine loses distinct awareness of both herself and the divine person, *and so does the divine person.*

Though Plotinus, as I said, does not affirm this equality in the merging, he does note that this supreme oneness of soul and the Divine is imitated by "lovers and beloveds here below wanting to unite" ( *VI*, 7, 34). But Plotinus does not explain further or develop this resemblance of the loftiest human experience to a sexual bodily uniting. Hadewijch goes on to describe her merging with God, in terms reminiscent of sexual uniting, as a yet further experienced equality with God. Hadewijch affirms here a certain mutual awareness and, if we take her literally, a certain mutual dependence. Before the complete merging, says Hadewijch, her Beloved and she receive each other. She concludes her account of Vision 12: "In that abyss I saw myself swallowed up. Then I received the certainty of being received [*ontfaen*], in this form, in my Beloved, and my Beloved also in me" (172f.). This mutual receiving is not with Christ merely as man. It is, for instance, the countenance of the Holy Spirit which says, "You will be love, as I am Love....In my unity, you have received me [*ontfingstu*] and I have received you" [*ontfaen*] (Vision 3, 1f.).

The mutual, aware receiving by Hadewijch and God of each other characterizes not only this climactic stage when Hadewijch has been taken out of the spirit and plunged deep into the divine abyss. It characterizes as well the prior stage of the fruition when she was taken up in the spirit. In this earlier union and fruition, Hadewijch does not yet pass away into the other but sees and listens. But in this stage, too, she and the Beloved receive each other. "So can the Beloved, with the loved one [*lief met lieue*], each wholly receive the other in all full satisfaction of the sight, the hearing, and the passing away of the one in the other (Vision 7, 64f.; so, too, Vision 3, 1ff., Stanzaic Poem 34, 49–54. Cf. Vision 11, 37ff.).

Is this mutual receiving a break from mainstream theology up to Hadewijch's time? Again we look to Bernard whose affirmations of mutuality in union with the divine comes close to those of Hadewijch and who influenced her extensively. If Bernard never affirms explicitly a mutual "receiving" by Christ and the soul, he implies it. He repeats often in epitomizing this union John's "Who abides in charity, abides in God and God in him" (e.g., SBO II, LXXI, 7–10). If by love the two come to abide in each other, then they must in some sense receive each other. Hadewijch, who also repeats the Johannine theme of mutual dwelling (e.g., Letter 9: *dat een lief in dat ander woent*) may have been merely more explicit in asserting mutual reception.

On the other hand, what theologian before Hadewijch described as the beginning of supreme human union with God that Christ "came to me humbly as anyone who wholly belongs to another?" (Vision 7, 64ff.). That sounds like Christ takes on a certain dependence on the human loved one. To assert that Christ approaches in humility the raptured soul as one who wholly belongs to the soul does not, I suggest, sound like anything theological before Hadewijch or during her time.

Hadewijch reports a more substantial mutuality: she and Christ in some respect affect equally each other. Thus in her supreme union Hadewijch and Christ are "wholly flowing through each other" ( Stanzaic Poem 4, 47). This mutuality is not merely with her Beloved. He is man as well as God, and the mutuality of Hadewijch and her Beloved, as affirmed in Stanzaic Poem 4, might be taken to pertain to him only in his humanity. In Vision 14 (77ff.), Hadewijch is less ambiguous. She declares herself to be one of those who have lived human and divine love in one single being "so as to have been flowed through by the whole Godhead, and to have become totally one, flowing back through the Godhead itself." Hadewijch says actually: "flowing through into the Godhead" ( *dore vloye in die godheit*), which reinforces the impression of Hadewijch affecting God. Any suspicion that Hadewijch means that Christ alone does this flowing is removed by later lines (145ff.) where she declares that when she is out of the spirit and in her Beloved she "is not less than he himself is" and "wholly such as he [is] who is our Love." This mutual flowing with God anticipates what happens between God and his friends who have attained eternal beatitude. God and his friends eternally "in mutual interpenetration [literally, "through-flowing" ( *dore vloyeke*)] enjoy such blissful fruition, and are flowing into his goodness and flowing out again in all good" (Letter 12, 53ff.).

In some passages where Hadewijch describes the supreme union of herself with Christ, her single word for the two of them suggests a mutuality of relative equals more strongly than Hart's translation by distinct phrases. Where Hart translates "the Beloved," i.e., Christ, and "the loved one," i.e., the human person, Hadewijch wrote usually the same word, *lief* or, in feminine gender, *lieue* or a corresponding pronoun (e.g., Vision 7, 64ff., Letter 9, 1ff. Stanzaic Poem 4, 47; cf. Hart 1980, 8). Hart's translation is defensible and useful, and I will at times use it, but a more literal translation brings out further how Hadewijch suggests a relative equality between the human loved one and the divine.

For instance, in Letter 9, Hart translates "with what wondrous sweetness the loved one and the Beloved dwell one in the other." But Hadewijch says: *dat een lief in dat ander woent*. This is more literally translated: "with what wondrous sweetness the one loved one dwells in the other." In following assertions of the letter, the one loved one and the other are so referred to, but not distinguished or separately identified. Similarly, Hart translates in Vision 7, "So can the Beloved, with the loved one, each wholly receive the other in all full satisfaction of the sight, the hearing, and the passing away of the one in the other." But "the Beloved, with the loved one" is, in Hadewijch's words. simply *lief met lieue*. Hadewijch, here as often elsewhere, does not by her words distinguish between Christ and the human lover in describing their relationship as lovers.

On the other hand, Hart's translation at times expresses a more active mutuality than Hadewijch does. Hart by translating in Letter 9, "and how they penetrate each other," expresses a rather active, phallic sense. So too, Van Mierlo in his paraphrasing note, *dore het doordringend* (1947, 79). But Hadewijch says: *Ende soe dore dat ander woent*, literally: "and so through the other dwells."

In Letter 9 Hadewijch also affirms a further mutuality of the two loved ones: "they abide in one another in fruition." By the three English words, "abide...in fruition," Hart translates one word: *ghebrukken*. Words with the *ghebruk-* stem are used by Hadewijch frequently to designate her supreme union with God. They are translated by Hart usually by "fruition" or phrases with that word. "Fruition of God," as we noted above, is what Hadewijch most prizes and most seeks. It is the single final goal for all humans, what they want, what God wants for them. It is what the blest have for all eternity. God grants it on occasion to chosen souls on earth.

Hart's translating of *ghebruk-* by "fruition" makes sense. The two words' stems, *bruc* and *fruc*, are connected, perhaps originally identical, in old European languages. Hadwewijch's use of *ghebruk-* words is roughly the same as common medieval theological usage in Latin of *frui* and *fruitio*, the meaning of which generally includes pleasure, but not exclusively and often not primarily. When Thomas Aquinas, writing at the same time or a little after Hadewijch, uses *frui* or *fruitio* to designate the supreme union of the soul with God, he means something like a "possession" or "use" ( *possessio, usus*) that, at least secondarily, entails or enables enjoyment or pleasure. Thomas's *fruitio* means (1) a having or using (2) that is pleasureable. It has thus a two-sided meaning though in a given context one side may be in the shadow or even not thought of (e.g., ST I–II, 2, 7; 3, 1–4; 4, 1–5; 5, 2–3; *In Joann.* c. 17, lect. 1, n. 3, ed. Marietti, p. 442a, quoted in ST I–II, p. 19).

The modern Dutch word stem, *geniet-*, with which Van Bladel-Spaapen and Mommaers translate *ghebruk-*, has a similar double meaning. So too, the English word "enjoy." One can "enjoy" rights or health even at a moment when one does not "take pleasure" in the rights or health. On the other hand, the word "enjoy" generally implies at least, that one will, if one becomes aware of them, take pleasure in the health, rights, or whatever is "enjoyed." Of *ghebruk-* words, "enjoy," "enjoyment," and so on are at least as faithful a translation as "fruition." It communicates better to the modern reader what Hadewijch says, for the context makes repeatedly evident that she has pleasurable union very much in mind.

Take the clause in Letter 9 where Hart translates, "They abide in one another in fruition, mouth in mouth." A more communicative as well as more literal translation would be, "They enjoy mutually each other, mouth in mouth" (*si ghebrukken onderlinghe ende ele anderen Mont in mont*). See also, e.g., Stanzaic Poem 34, 49ff.

That Hadewijch has pleasuring in mind is evidenced by her affirming the "sweetness" (*soeteleke*) of the mutual indwelling of the two loved ones as well as "the sweetness of the divine nature" (*ene soete godlike nature*) flowing through them. A typical passage is: "But how much sweetness is found in the interior feeling [*gheuoelne*] and fruition of the Beloved [*ghebrukene van lieue*], all these who were ever born in the human shape could not fully explain to you" (Letter 27, 44ff.). Hart does at times use "enjoy" to translate *ghebruk-*, e.g., in the passage of Letter 12 (53ff.) cited above concerning the mutual flow of God and creature in eternal beatitude: "God and his friends...enjoy such blissful fruition [*weeldeleke ghebrukende*]."

Their indwelling is, therefore, mutually sweet to Christ and to Hadewijch. He, as well as she, enjoy each other as they hold each other. "The *lief* and the *lieue* embrace each other / And have fruition in giving themselves to each other" (Stanzaic Poem 34, 54). They anticipate what God and his friends do in the afterlife. As they give themselves to each other, they affect each other, give each other pleasure. Correspondingly, they are affected by and receive from each other.

Before culling and considering assertions by Hadewijch of an even more substantial mutuality in her supreme union with God, it is worthwhile to pause and ask: In what respect, if any, is the mutuality with God that we have so far heard Hadewijch affirm new to Christianity? Mutuality between God and humans has been a common feature of faith and devotion of Christianity since its origins. It was a common feature of the religion of ancient Israel and even, though less so, of Greek and Roman popular religion. The New Testament continues the Old: not only does God do many things to humans devoted to him. They do many things to him: please, displease, anger, sadden, appease, and so on. True, as we will discuss in the next chapter, the Christian dogma of the complete self-sufficiency of God and his conse- quent complete immutability—dogma heavily in debt to Greek and Roman philosophy—made it difficult to explain how any creature could affect God. But the dogma did not prevent theologians since the beginning of Christianity from holding that God was pleased by per- sons faithful to him, displeased by sinners, and so on, while debating, century after century, how this was possible. The God of Abraham, Isaac, Jacob, and Jesus Christ interacted with his people and, in some sense, showed himself affected by them, as any loving ruler or father is by his subjects or children.To imagine, however, the mutual love of God and creature as two lovers erotically enjoying each other was a revolutionary step in Western religious thought. It pictured an intensely personal mutual interdependence and interaffectivity.

Hadewijch was far from the first to take the step. The pagan Plot- inus, creator of the Neoplatonism that flooded Christian thought, while insisting on the inexpressibility of the soul's supreme union with highest divinity and the unqualified dependence of all on the self-sufficient, solitary One, did observe, as we saw, that this union was like that of lovers. But he did not say in what the resemblance con- sisted; in his writing he stressed the difference (VI, 7, 34; 9, 3, 9). Fol- lowing Plotinus, Pseudo-Dionysius affirmed that the One can be called "Desire" (*eros*), for it "desires all things." But he explained that what "touches" and "transports" *eros* outside itself is simply Itself as Love, Desire, and Good (*The Divine Names* IV, 10–15).

Another erotic influence from a closer source converged on Christian thought about the supreme union of soul with God. Christ- ian theologians from Origen on often interpreted relations between God and the soul after the model of the two "spouses" of the Song of Songs. That book of the Bible, after the Psalms, seems to have received the most commentaries in the medieval Christian West (Origen 1966, 27). It was "the book most read and frequently commented on in the medieval cloister" (Leclercq 1960, 106).

In the early twelfth century, Bernard of Clairvaux, inspiring many imitators in that century as afterward, wrote numerous sermons on the Song of Songs. He applied at length and in detail the biblical account of the two lovers to the relations between God, especially the Divine Word, and the devout soul. This relationship, in general and not only on its occasional mystical summit, was described by Bernard as in the book of the Bible, as being mutual, transpiring between two persons giving and taking pleasure to and from each other as relative equals. In bringing out the soul's contribution to the embrace, Bernard does not hesitate to say that a human soul can by its love please singularly the Divine Word, prove itself worthy of his loving embrace, contribute to the serenity of the Divine Visitor, give him embraces and kisses which he seeks and takes pleasure in, has drawn God into itself by its returning love (SBO I, VIII, 7; XXIII, 1; XXIII, 15–16; XXXI, 6–7; XXXII, 2; SBO II, LXXI, 10; LXXII, 3).

Commenting on "Let him kiss me with the kiss of his mouth" in the Song of Songs, Bernard says no sweeter words can express "the sweet feelings of Word and soul for each other (*Verbi animaeque dulces ad invicem...affectus*)" than "bridegroom" and "bride" (SBO I, VII, 1: SBO I, 31–32). Not only does God give the devout soul its holy pleasures and joys. The devout soul gives God pleasure: it sings often festal songs that "stroke pleasingly the ears of God" (*divinas mulceat aures*) (SBO I, 9; cf. SBO II, LI, 4, *beneplacitum sponsi*).

Bernard, however, renders the same philosophical explanation of this mutual love and pleasuring of God and soul as preceding theologians gave generally of the relations between humans and God. Bernard maintains the complete independence and immutability of God. God could in no real sense depend on anyone who is not God. Whatever be the image or phrase used, God could, in fact, receive nothing from or be in no way affected by the soul. He thus indicates a different experience from that of Hadewijch as well as a different ideal for human life.

How then do the Word and the devout soul speak lovingly to each other? Not by their bodily word nor their bodily appearance but by their spiritual speech. The speech of the Word is His gracious favoring of the soul. The speech of the soul is its fervent devotion, brought about by the Word's favoring.

> For the Word to say to the soul, "You are beautiful" and to call her "friend" is to infuse the soul that by which she may both love and presume she is loved. For her in turn to call him "beloved" and confess that he is "beautiful" is to ascribe sincerely to him that he loves and is loved and to marvel at his favoring and to be stunned by his grace. (SBO II, XLV, 7–8; similarly SBO II, LXVII, 8)

The speech of the Word is the infusion of the gift; the response of the soul is grateful admiration. She loves, then, even more because she senses herself conquered in love (XLV, 8).

Generally speaking, therefore, the devout soul interacts with the Word in that the soul experiences love for the Word surging in her. She presumes rightly that her growing love is a gift from her divine lover. She hence loves him all the more. The experience is intimate and profound. It is not truly known except by those who have the experience.

But the mutuality thus experienced, as Bernard theologically explains it, is limited, far short of what loving human mutuality would be—all the more if erotic.

First, the soul does not experience directly the Word nor his love itself. She experiences only herself loving more. She knows, by inference from her own love, that the Word loves her greatly. She, therefore, loves him yet more, still without experiencing him. Secondly, throughout this intimate exchange, as Bernard explains it metaphysically, the Word does not love her *because* she loves him. Rather she loves him because he loves her. The Word's love is purely outgoing, giving, not receiving (XLV, 8–10; LXVII, 8–10). As noted above, Bernard affirms with John that God is in the loving soul and the soul is in God. But Bernard explains both indwellings as God's action on the soul: respectively, creative, continuing conservation of all souls and infusion of love into souls that love God. The soul does not act on God, does not affect God, and gives nothing to God.

What we have just heard Bernard say, he says of the soul's relations in general with God. He says the same specifically of his own mystical experience. Bernard of Clairvaux at times disclaimed humbly that he had had any mystical experience. At other times, he humbly acknowledged and described his mystical experience (LXXIV, 5). He revealed here basic differences from Hadewijch's experiences.

First, as already noted, for Bernard the supreme union with Christ did not take bodily form. Bernard emphatically denied bodiliness here (LXXIV, 5; XLV, 7; LII, 5). He readily imitated Scripture which imparts to us "the incomprehensible and invisible things of God by means of figures drawn from the likenesses of things familiar to us, like precious draughts in vessels of cheap earthenware" (LXXIV, 2). Bernard knew that Christ does appear to select souls in bodily form, visible and audible. God thus through the body communicates truth and wisdom to such souls. But Bernard not only disavowed having such "visions." He disparaged such experience in contrast to the highest earthly experience of men, the supreme union with God, to which God at blessed moments raised him and other select souls.

This experience was purely spiritual. It had nothing bodily in it. Bernard's description of that union, even his sensuous, passionate evocation of it in sermons on the Song of Songs, has to be intended as only metaphor (XXXI, 6; LII, 5; LII, 3, LXXIV, 5–6, LXXXV, 13). What happened was similar but not the same as what he described. Hadewijch, on the other hand, is, in passages such as Visions 6 and 7, clearly not intending metaphor. She wants to describe directly what happened.

Its disembodied nature, I argue later, is one reason why in describing his own experience of supreme union, Bernard denied, or at least omitted, any direct experience of mutuality between God and him. For one reading Bernard after reading Hadewijch's report of her visions, this omission is striking. It is even more striking and curious in light of the mutuality he attributes to such sublime moments when he preaches on the Song of Songs. Bernard, for instance, did not, as Hadewijch does, report that he was conscious of receiving the Beloved as the Beloved received him. Bernard, on the contrary, acknowledged that he did not know the moment when the Word comes to him (LXXIV, 5; cf. 1). He realized eventually that the Word had indeed come and embraced him. However, Bernard did not experience this embrace directly but inferred it from his own extraordinarily swelling, rapturous love (LXXIV, 5-7). The Word must be present, in that he infused wondrously this divine love into Bernard. This is all happy, certain inference, but he does not experience the Other himself.[1] Bernard does say he experiences the divine nature, i.e., divine love, but he explains this as his experience of his own love growing mightily until it becomes like divine love insofar as it can be and is aware of and loves nothing but love.

What Bernard experiences directly, by this account of his, is his own love, i.e., his mounting desire, pleasure, and satisfaction. What he does not experience directly is the other person. He does not experience the other person being affected by him. He does not experience himself affecting the other person. This analysis of a somewhat masturbatory ecstasy falls well short of the erotic imagery of the Song of Songs, which Bernard applies often elsewhere both to Christian life in general and to his mystical experience. There, as we saw, he declares that both persons relate to each other in mutual seeking, mutual embrace and mutual enjoyment.

By his theological explanation of the soul's general relations with God and of his own mystical experience Bernard not only turns into mere metaphor, or simply contradicts, his account of the supreme mutuality of soul and the Word on earth. He also, all the more, falls short of the kind of mutuality we have heard Hadewijch describe in her own experience. He excludes the possibility that he was, in fact, consciously experiencing: (1) himself and the Word melting indistinguishably into each other, (2) their receiving of each other, (3) their flowing back and forth through each other, or (4) their "sweet" giving themselves to each other. Bernard excludes these possibilities because he does not envisage his soul experiencing the Word in this life. He excludes these possibilities because the Divine Word cannot be affected by or receive from a human being.

It is evident that Hadewijch has come a long way from Augustine and Pseudo-Dionysius in affirming mutuality as constitutive of the best loving available to humans on earth. Not so evident to me, despite the differences I have just traced, is how far Hadewijch has come from Bernard of Clairvaux and twelfth-century theologians who follow his lead. I have not combed the works of Bernard thoroughly enough to exclude that in other passages he affirms and explains philosophically a greater mutuality between God and the soul in the soul's supreme loving than I have so far seen. This present book is a report on work in progress and an invitation to coworkers.

Should one make much of what Bernard says speaking as philosopher about God when it contradicts what he says interpreting the Bible. Those of us may make something of it who believe there is truth in Hegel's identification of philosophy as "the mind of the times (*Zeitgeist*), grasped in concepts. After all, Hadewijch, too, is a philosopher and at times articulates philosophically her supreme union with God. She articulates it differently than Bernard. Perhaps in Bernard we witness a struggle of that *Zeitgeist* and in Hadewijch a resolution.

In any case, my primary concern in contrasting Hadewijch with Bernard is not to gather evidence of Hadewijch's originality. Much more evidence is needed, e.g., by comparing Hadewijch with other twelfth-century theologians such as William of Saint Thierry or the Victorines. My primary purpose in contrasting Hadewijch with Bernard is to bring out more sharply Hadewijch's own answer to the millennial Western question: What is the best human living possible? In the following chapter, therefore, we return to the words of Hadewijch and hear her affirm, more unequivocally and systematically than we have heard her affirm so far, *how* God and she are present to each other, affect each other, depend on each other, give to each other, and receive from each other. These affirmations constitute a substantial, provocative answer to that millennial question. These affirmations are all the more thought-provoking for their resemblance with and extension of replies by late twentieth-century women thinkers to this same question.

## 3

In the supreme embrace that is the human individual's supreme knowing and loving, God and the human individual not only enjoy each other. They not only take pleasure in each other and give pleasure to each other. They also content and satisfy each other. In the words of Hadewijch, they "are enough" (*ghenoech sine* or *ware*) for each other. It is not enough for each of them until they are united. Sometimes she says they "do enough" (*ghenoech doghene*) to each other. But more often, she says one "is enough" (*ghenoech sine*) for the other. In Hadewijch's usage, the phrase *ghenoech sine* or *doghene* connotes usually "to give pleasure" but denotes more than that, as her words in Vision 7 illustrate:

> My heart and my veins and all my limbs trembled and quivered with eager desire and, as often occurred with me, such madness and fear beset my mind that it seemed to me I did not content [*ware ghenoech*] my Beloved, and that my Beloved did not fulfil my desire....I desired to have full fruition of my Beloved, and to understand and taste him to the full. I desired that his Humanity should to the fullest extent be one in fruition with my humanity, and that mine then should hold its stand and be strong enough to enter into perfection until I content him [*ghenoech ware*], who is perfection itself, by purity and unity, and in all things to content him [*ghenoech te doghene*] fully in every virtue. To that end I wished he might content me [*ghenoech ware*] interiorly with his Godhead, in one spirit, and that for me he should be all that he is, without withholding anything from me....

> ...to let everything come and go without grief, and in this way to experience nothing else but sweet love, embraces and kisses. In this sense I desired that God give himself to me that I might content him [*ghenoech te sine*]. (1ff.)

Hadewijch sums it up toward the end of this vision, as quoted in the preceding chapter: "So can the Beloved, with the loved one [*lief met lieue*], each wholly receive the other in all full satisfaction [*ghenoechten*] of the sight, the hearing, and the passing away of the one in the other" (64ff; see also Letter 7).

What gives pause is not that Hadewijch pleases God. It is that she contents and satisfies him. She becomes somehow enough for him. Not all pleasure or enjoyment contents and satisfies. Not all pleasure or enjoyment is enough. One can be pleased with something, enjoy it, and not feel contented or satisfied with it. It is not enough. One desires more from it. One's present desire is for more pleasure or enjoyment, from it or from something else. It is not yet enough. I may enjoy a particular meeting with a particular person but not feel contented or satisfied. It is not yet enough for me. Birdsong outside my window may give me pleasure this morning, yet it may not (or may) satisfy or content me.

That human subjects can please the Divine Ruler, give pleasure to him, is traditional theology from the origin of Christianity to the present day. The theologians accept it as a given of Christian faith and wrestle with problems that the dogma poses for their unquestioned axioms of divine self-sufficiency, immutability, infinity, and so on. That a human subject can content and satisfy the Ruler, be enough for him, says more than giving him pleasure but, in certain senses, is affirmed by the Christian theological tradition. Humans are seen in some sense to content and satisfy the Divine King by "satisfying" his various wills for them: his laws, callings, judgments of punishment, forgiving mercy, and so on.

But traditional theologians do not conceive that simply to be loved by a human individual can "satisfy" some part of God's own personal life. They do not conceive that my love for God, and only that love, can be *enough* for God in some part of his inner life. The theologians cannot conceive this for they cannot concede that God by himself in his inner life is not completely enough for himself, is not completely self-sufficient, does not completely satisfy and content himself. They cannot, therefore, conceive that the supreme, final union of God and the human individual, the ultimate goal and norm for all human life, includes a mutual satisfying by God and the human being, a unique contenting of each by the other. But Hadewijch experiences and then conceives exactly that.

In affirming the mutual *ghenoech sine* of God and human person in intimate embrace, Hadewijch means that each is enough for the other's *desire* for the loving embrace. We see it for the human in Vision 7 above. We will see it for both God and human in passages soon to be quoted. In these passages, the comparison of Hadewijch is not with a benign ruler and subject, with a caring father and errant son, with a shepherd and lost sheep. The father and shepherd of Jesus' words do desire to hold the son and sheep, but they desire it essentially for the good of the son and sheep who are presently lost. The desire of the ruler, too, is for the subject's good. Correspondingly, it is for their own good that the subject, son, and sheep turn to the superior other. But Hadewijch compares God and soul with two persons who desire simply their delightful, passionate embrace. The two may be erotic lovers or friends or mother and child. Their embrace contents and satisfies both God and the soul simply because they both desire it for its own sake.

To satisfy desire is a common meaning of the phrase *ghenoech sine* in medieval Dutch, as in the corresponding Latin, *satisfacere*, which etymologically means "to make or do enough for." That Hadewijch means that God and a human lover can be enough for each other's desire is stated in numerous passages such as those I have quoted and am about to quote. This meaning is reinforced in some of the passages, as we will see, by metaphors of satisfying hunger and thirst. Hart's more common translation of *ghenoech sine* or *ware* by "to content" is correct, but "to satisfy" in most contexts would have brought out more sharply what Hadewijch is saying: to be enough for the other's *strong* desire. In English one speaks rather of "satisfying" than "contenting" one's passion, hunger, or thirst. In any case, Hart does at times translate Hadewijch as affirming that the soul satisfies God (e.g., Stanzaic Poems 4, 5; 30, 82; 34, 5).

That the human individual in mutual enjoyment with God can "content" or "satisfy" (*ghebruk-*) God does not keep Hadewijch from maintaining, as medieval thought generally did, that God always enjoys himself, has his own "fruition" of himself (Letter 12, 53ff.; Letter 6, 19ff.; Letter 22, 102ff.; Letter 21, 1ff.; cf. ST I–II, 2-5). "What he is, he lives by, in his sweet self-enjoyment" [*hi selve in siere sueter ghebrukenissen*] (Letter 1, 56ff.). Hadewijch affirms that God, his "sweet nature," Love, contents, satisfies, is enough (*ghenoech*) for himself (Letter 22, 31ff., 348ff.; Letter 2, 66ff.). For Hadewijch, therefore, to a certain extent, God is self-sufficient.

Hadewijch points out, at times bitterly, at times trustingly, how those who love God on earth wander in painful privation of fruition while God has fruition of himself to the full (e.g., Letter 1, 41ff. and 56ff.; Letter 2, 51ff.; Letter 6, 19ff.; Letter 12, 53ff.; Vision 6, 92ff.). Repeatedly through her writing, Hadewijch laments the common lot of the lover of God who "stands in the chains of love in nonfruition and disgrace" (Stanzaic Poem 9, 53–56). But God enables and invites human beings not only to serve Divine Love through their earthly life but also, at least after death, to "enjoy" (*ghebrukene*) Love's being where Love loves herself, enjoys herself, and "satisfies herself" (*ghenoech met es*) (Letter 2, 66ff.; Letter 20, 123ff.; Letter 22, 25–46; cf. Vision 6, 67ff., 76ff.; Stanzaic Poem 2, 31–36; Stanzaic Poem 15, 21–24). The human individual's fruition of God is but a share of God's, Love's, satisfying fruition of himself for all eternity (Letter 6, 19ff.; Letter 2, 66ff.). The individual has fruition of God in that he or she loves God with God's own love (Letter 16, 9ff.).

This is why, by her fruition of her Beloved, Hadewijch knows "how in fruition he embraces himself" (Vision 10, 145ff.). But God does not enjoy only himself. He has, too, eternal fruition of his creatures. He has some fruition of the individual soul while the soul wanders in privation of that divine fruition (Letter 1, 33ff., 41ff.; Letter 2, 51ff.; Letter 19, 37ff.; Letter 22, 102ff.). But the fruition that he has of every soul wandering on earth is not enough for him. He desires the soul further. He "wants a total fruition of her" (Poem in Couplets 12, 51–52). If the soul put all else out of mind save the unity of Love, then, as in Psalm 45:11, "The King shall desire your beauty."

> Of which he with his whole Nature
> Wishes to have fruition in one permanence—
> And that beauty will meet with one Beauty,
> And they will greet with one single greeting.
> And that kiss will be with one single mouth. (83–87)

The soul, therefore, satisfies God in their supreme union because without that union God has unsatisfied desire to have this fruition of the soul. The blest living with him in heaven do satisfy this desire of his in their regard. They are eternally engaged in contenting him. They content God by loving him with his own love, becoming one with him in one love. This is their ever incomplete, completely blissful fruition of God. But those on earth for all their striving cannot content their "never-contented Beloved." They know it (Letter 16, 14ff.).

Still, on exceptional occasion, momentarily, they do. Even on earth chosen individuals at privileged moments content God. How? Not by the person's usual life, though if a person has not lived a loving, virtuous life, he or she cannot move on to the supreme fruition in which he or she contents God. In other contexts Hadewijch urges in general a loving life in serving, suffering, seeking, and so on, but not necessarily a supreme union, on the grounds that this common life of love does content God (e.g., Letter 12, 13ff.; Stanzaic Poem 8, 36–42). But what contents and satisfies God in that exceptional moment of supreme union is simply the person's present loving union with God.

When "the soul thinks of nothing else but kissing him and being within him, this is God's life and pleasure." He then adorns the soul with love according to his pleasure (Poem in Couplets 12, 63–68; my translation). Traditional theologians would fault Hadewijch for her illogicality. If God suffices for himself, satisfies himself, then nothing else, no one else, can be said to suffice for him or satisfy him. But the issue, I suggest, is not one of logic. The issue—yes, the *rational* issue— is of rock–bottom judgments on which all one's intellectual structure is built. Aristotle calls such judgments *archai*. Thomas Aquinas calls them *prima principia*. Moderns call them faiths, intuitions, categorical imperatives, constructions, or other terms.

The issue of rock-bottom judgment which Hadewijch pursues concerns what good personal life consists in. Her judgment is that such life, for God or humans, is a sufficiency in self *and* a sufficiency in others. Conceptually, there is no contradiction here. One can propose: "In one respect...yet in another respect..." Experientially, I submit, it is commonplace. In a good, live, mutual love or close friendship, I *feel* self-sufficient and self-satisfying in myself and also just as profoundly sufficient and satisfied in my loved one, my friend. Correspondingly, Hadewijch lays out that the best mutuality does not arise from *sheer* longing. The best mutuality arises out of a certain self-sufficiency of each person in themselves. In this interconnection of personal values, *absolute* self-sufficiency has no place. But it is out of and then with a *relative* self-suficiency that each mutually desires the other. This banal, precious experience is, of course, inconceivable to Plato and his footnoters.

It is already to some extent clear how for Hadewijch God suffices in himself. His fruition of his very being, his nature of Love, his loving—three different conceivings of the same divine reality and life— satisfies him. He does not *desire* himself nor fruition of himself because he eternally has himself and fruition of himself. But he does at times desire individual souls, and he desires them and the fruition of them. Hadewijch tells us something more of her experience of this.

God gains not only pleasure and satisfaction from the soul, as does the soul from him. He gains a good for himself as the soul does for itself. In this union with the soul God gains his liberty, as the soul gains its. Any soul, if it maintains its worthy state,

> is a bottomless abyss in which God *contents* himself; and his own *contentedness* ever finds fruition to the full in this soul, as the soul, for its part, ever does in him. Soul is a way for the passage of God from his depths into his liberty; and God is a way for the passage of the soul into its liberty, that is, into his inmost depths, which cannot be touched except by the soul's abyss. (Letter 18, 63ff.)[1]

There is something "touching" in how God and soul each find their own freedom in the other's depths. Touching and familiar. One may recall one's own experience. Note that whereas Hadewijch spoke elsewhere of the soul and God contenting or satisfying each other, here she speaks of each contenting or satisfying themselves. They fit together if we reflect on the ordinary human phenomenon of good mutuality. They also constitute an interesting conceptual connection: it is in freedom with another that one satisfies one's desire.

Hadewijch, in my reading, does not identify further this "freedom" gained by God and soul in their union. It is perhaps the free choice that they each make to carry out the drive of their desire. We will hear Hadewijch below affirm exultantly that she made this free act of will (Vision 11, 98ff.). She speaks regularly of God's or Love's "will" and it often refers to God's decision to act this or that way (see Hart's Index under "will"). In any case, from the text itself this "freedom" is something precious, distinct from their nature, something that both the soul and God want and will very much to get but can get only in union with each other. Their union, this summit of all human living, contains, therefore, in their very freedom some real *receptivity*.

On the other hand, it is "driven," "flung," by divine nature that Love goes forth and gathers in the human seeking mind, the desiring heart, and the loving soul, and casts them into the abyss of her divine nature and has therein fruition of her nature (Letter 20, 1ff.). God thus wills to satisfy what is not any desire but desire felt as overwhelming force of herself on herself. So too, the soul even when it can exercise free will and refuse the drive (Vision 11, 98ff.). I would call this desire "need" and argue its trueness to experience of good mutuality though "need" is not a popular word with contemporary champions of mutuality whom I have read.

Carter Heyward uses the word and synonyms to stress that we need each other, but mostly she means a need of friends and lovers so that we have power in our further action and living (1989, e.g., 56, 94, 98–99, 105). I have found no passage where she unequivocally affirms that we *need* or *depend on* each other just to be together, pleasuring and satisfying each other by ourselves, though her words seem to imply it (cf. 14, 21–22, 33–34, 100, 104, 112, 122, 132). I find the same, to me surprising, reticence and ambiguity in Mary Hunt's excellent *Fierce Tenderness* (1991) and other feminist works. Perhaps still burning from the traditional male identification of woman with overpowering desire, contemporary feminists hesitate to stress the reality and force of spontaneous desire in mutuality and do not fully resist the temptation to center the heart of mutuality in two relating but absolutely self-sufficient selves.

Hadewijch, on the contrary, affirms emphatically both the compelling desire and the supreme freedom of ideal lovers. She seems to me to affirm here a certain "passivity" of the lover to himself or herself, a passivity of God and herself each to their own pressing desire. It is a passivity that will be integrated, mastered perhaps, by freedom of will. But the passivity itself, the rising, nature and force of the pressing desire, is not subject to will but given to it to carry out or deny.

How strong is the desire that spontaneously moves God and soul each out of themselves to unite with the other, Hadewijch expresses also by metaphors of hunger and thirst: "...we shall live with one life, and one love shall satisfy the hunger of us both" (Letter 31, 1ff.). The "intimate exchange of love" between God and her is "the custom of friends between themselves to hide little and reveal much [so that] what is most experienced is the close feeling of one another, when they relish, devour, drink and swallow up each other" (Letter 11, 10ff.; see also Vision 1, 391ff.). In the union of the loving soul with Love, a union with one will and one being, the "depth of desire pours out continually, / And Love drinks all that outpouring" (Stanzaic Poem 12, 23–30).

In substantial aspects, therefore, the description by Hadewijch of supreme mutuality of divine and human persons breaks from the traditional account of that union, even from the account by Bernard of Clairvaux. Bernard made similar comparisons of God and the soul to lovers, but he did not say that the soul "contented" discontented Divine Love nor that the soul "satisfied" God. As we saw, Bernard's God desires souls. He wants souls to love him. He wants only that from them. But he wants their love "knowing that they who love him are blest by their love itself."[2] For Bernard, God desires the soul not that God may get anything from the soul. God desires the soul simply so that he may give love to it. For Hadewijch, God wants the soul, yes, so that he may give love to it, but also so that he may get something from it. He wants to have fruition of the loving soul. He wants to satisfy his desire for the soul. He wants to obtain therein a new liberty for himself.

Bernard does say that when the bride abounds in love, the bridgegroom is "content" (*contentus*) with that love. His desire ceases. He "seeks" (*quaerit*) nothing else. These words may have stimulated Hadewijch to her further development of the theme. But Bernard does not say that the loving soul thus *contents* the bridegroom. In Bernard's theological explanation, God contents himself. He does so simply by giving the soul love. He does not seek anything for himself, as indeed he gets nothing for himself. God seeks only that "they who love him [be] blest by their love itself.' Bernard the theologian does not take seriously the erotic analogy proclaimed by Bernard the preacher.

Nor does he take seriously the analogy with eating and drinking. In Sermon 71 on the Song of Songs, Bernard asserts that God and we humans eat each other. But in the eating he refreshes us with his joy while he rejoices at our spiritual progress. Bernard does not say that we give God joy or that we satisfy his desire. God's food is our penance, our salvation. He chews, swallows, digests, and assimilates us in that he rebukes, teaches, changes, transforms, and conforms us. Thus, are we more closely and strongly bound to him, perfectly united to him, truly in him as he in us. But Bernard does not say that this eating satisfies any desire or hunger of God. No appetite presses him, as the Divine Love does God, according to Hadewijch. The mutual eating, as Bernard explains it, is not really mutual: God gives all and receives nothing. We give nothing and receive all.[3]

Hadewijch says what Bernard says, but she leaves Bernard behind when she sees herself satisfying God, affecting God by her passion and will. I have found no other Western Christian theologian contemporary with Hadewijch or before who affirms that human beings satisfy or in any way affect God in their supreme union with him. I want to say " *actively* affect" God, but "active" is a traditional western word expressing a traditional Western concept and the concept does not fit Hadewijch's "satisfying" of God. In a passionate embrace of God she is not being strictly "active." In traditional Western philosophy and theology, i.e., one "acts" on, "does" something to another, only in giving what one already has. Hadewijch does not already have the satisfying pleasure and liberty God gains in the embrace but rather gains her own at the same time as God gains his. Indeed, what of Hadewijch we translate as "satisfy" or "content" is usually, literally, not "do enough" (*ghenoech doghene*) to each other but "be enough" (*ghenoech sine*) for each other.

Similarly, I want to say that Hadewijch in her embrace "causes" in God satisfying pleasure and liberty but again, the word as used in traditional Western thought is a poor fit. From Plato on, to cause is to bring about a participation of what one has. An "efficient cause" does it by acting, as we just saw. A "final cause" does it by attracting love by the cause's goodness. If I may contort Latin and English: to "cause" for Thomas Aquinas is to "flow into" (*influere* as a transitive verb) something of what the cause, whether final or efficient, has. Hadewijch "flows into" (*influere* as an intransitive verb) God, affecting him with something she does not give him. God's being affected by Hadewijch is ultimately his fruition, his enjoyment of her. And God's enjoyment of Hadewijch is not primarily a communication of Hadewijch's pleasure. The pleasure God derives from their union is not, at least not primarily, a participation in Hadewijch's pleasure in him.

Yet the affecting of God by Hadewijch, and vice versa, is readily understood and appreciated by the reader. It is obviously analogous to commonplace happenings of human life. Imagine the lovers as in Vision 14. holding each other in fierce embrace. Imagine close friends embracing after long separation. Imagine a parent holding a sleeping child. Imagine that the father of the prodigal son turned once to the older son and held him tight in joy at having him with him now as always. A recognition of this mutual affecting by ones who love each other is found in traditional Western expositions of friendship such as that by Aristotle in the *Nicomachean Ethics*. But this kind of affecting by one person of another is not, for the tradition, one of the basic relations structuring all reality. God does not affect creatures in this way, nor—much less!—vice versa.

In brief, I know no theologian before Hadewijch who articulates systematically how something like the mutual satisfying and affecting of those who love each other and are at the moment together constitutes a supreme union with God, is the goal and norm of all human striving. I say no theologian. From the eleventh century on in the West, countless Christians, especially women, experienced something similar. Many a woman in an ecstatic vision or devout fantasy embraced the Divine Child while the Child embraced her. Many women cared for, played with, suckled the Baby. They contented the Baby while the Baby's pleasure satisfied them completely. Similar, too, is the widespread identification at the same time, in particular through the multiplying Pietàs, with the mother grieving with her dead Son, holding him as his weight presses down on her.

We will return in Part II to this new popular devotion of the Middle Ages. It is as revelatory as Hadewijch's writings of an ideal of supreme union of embodied mutuality. We will study facets of this supreme value of the experience of the women. We will focus particularly on the bodiliness of the union. In what remains of Part I, we ask further *how* Hadewijch and God affect each other. Chapter 4 dwells further on the contrast, begun in Chapter 3, between this affecting of God, as we have so far heard Hadewijch describe it, and traditional concepts of acting on and causing something in another person. Chapter 5 centers once more on Hadewijch herself: what she reveals further of her way of affecting God. She brings out how her affecting of God in her supreme union reflects her "strength," makes her "proud," and so on.

# 4

In the next chapter we will hear Hadewijch describe even more boldly how she affects God in entering supreme union with him. She tells that she "conquers" God while he conquers her. She describes her experience of this mutual conquest. Thus, she discloses further how she understands the mutuality that, for her, is the summit and summation of good, human life. She also, I suggest, raises difficult but fruitful questions for late twentieth-century thinkers, including feminists, though it lies beyond the limits of this book to develop these questions.

But the present chapter is a detour. To put into sharper relief the radicalness of Hadewijch's understanding of supreme mutuality, I offer here a more extensive and detailed meditation of the opposing human ideal, the aim, norm, and model for human life proposed unfailingly by traditional Western thought up to, during, and (I suggest) after the time of Hadewijch down to the present day. While repeating things I have already asserted and argued, I interconnect and advance them. I ambition in this chapter a spiraling that descends further into the matter. The reader impatient to go on listening to Hadewijch can omit this chapter without losing the sequence of thought of the book.

Let me recapitulate. The systematic assertion by Hadewijch that God and she in their supreme union of knowing and loving content each other and satisfy their mutual hunger, is odd—to my knowledge, unique—in Christian theology from its origins to the thirteenth century. Western thinkers, even twelfth-century theologians and story-tellers of courtly love, when they talk theology, speak Greek. No creature can content the Platonic Good, the Aristotelian "thinking of thinking," the Stoic "Logos," the Plotinian One, for such divine reality is completely self-sufficient. It has no discontent, no unsatisfied desire. So too, the Blessed Trinity. The God of the Christian theologians is wholly self-sufficient beause he is infinite being. Infinity as divine attribute begins in Western thought only with Philo. But it says no

more than earlier Western thinkers said. Supreme divinity is the fullness of reality. It contains all that is positive in all existing things. God,
therefore, can gain nothing because he lacks nothing. Nothing can be
added to it. No other reality can give it anything. No other reality can
affect it, for that would entail something happening to God in his personal life. Something then would be added to his being

Early Christian theologians affirmed in certain respects an interactivity between God and man that a Greek thinker would have
rejected out of hand. Think only of God's being affected by and
responding to the prayers of believers. Whether the affirmations were
logically consistent with the ongoing Greek theses of Christian
thought is not a concern of the present essay. But we find no mutual
affecting in early and Christian theological expositions of the supreme
knowing, loving union of God and soul. In that union, God affects the
soul profoundly, wholly. The soul does not affect God at all. With these
one-way expositions, the Greek thinker would have been comfortable.

As we saw above, in the ascent of Augustine and Monica and in
the soul's union with Pseudo-Dionysius's One, everything that happens happens to the soul. At this moment of union with God nothing
happens to God. It has to be this way. God (read: "the One" or "Logos"
or "thinking of thinking") created freely and would have been just as
contented if he had not. He would have been just as contented if he
had not drawn the soul to union with him. The union adds nothing to
his life. It cannot add to his pleasure or liberty, for they are already infinite. Nothing can be added to infinity.

The Reformers damned earlier Christian theology for attributing
too active a role to the human being. According to pre-Reformation
theology, individuals choose freely their salvation or damnation. They
merit whichever they choose. Hadewijch follows the tradition when
she affirms that her earthly, mystical fruition of God, anticipation of
her eventual eternal beatitude, is made possible by her free choice and
just deserts. Though for her and traditional Christian theologians, this
choice and merit is due entirely to God's gracious empowering, it is
also due to the human individual's free will. It is the work, act, and
contribution of the creature, too, though God gets nothing from it.

Whatever affecting of God by human beings that the theologians
inconsistently acknowledged in explaining human meritorious action
or divine response to prayer, they affirm no such affecting in humanity's final goal, the good because of which all else is good: the beatific
union in heaven or the mystical union that is the partial but closest
anticipation of the beatific on earth. There is no mutual affecting here,
and this supreme union must contain all intrinsic human values which
the theologians recognized.

God cares, of course, about the individuals who come to Him. He loves them. He wants them all to be saved. But he is not less happy, content, or satisfied if some choose damnation or choose not to respond to his mystical graces. In mystical union, as in eternal beatitude, God welcomes the soul lovingly into his arms. He rejoices in the union, is pleased. But—for the theologians, somehow!—the soul gives him really nothing. He gains nothing in the soul's embrace. He gains no more pleasure, no more happiness than he would have had otherwise, for his pleasure and happiness, like everything divine, is infinite and cannot be added to. This is surely an odd representation of the recorded love of the God of Jesus Christ, not to mention that of the God of Abraham, Isaac, and Jacob.

On the other hand, there is little in the Bible of the individual soul's experience of intimate union with God, whether for a moment in earthly life or for all eternity. It exceeds the scope of this book to analyze the few biblical texts that may do so, but I find nothing there of the mutuality of soul and God that Hadewijch experiences. Christian theology continued Greek thought as it evolved it. This is epitomized by the synthesis with which Thomas Aquinas, around the time of Hadewijch, wove the pattern of Christian belief and tradition into Aristotelian, Platonic, and Stoic stuff. More consistently than Bernard of Clairvaux, Thomas described both supreme mystical experience and eternal human beatitude with no inference, implication, or ambiguity that the blest person might here content, satisfy, or in any way affect God (e.g., ST I–II, 1–5; II–II, 175).

This confident description by early and medieval Christian theologians of supreme union with God will be maintained by many, but not all, Christian thinkers in the centuries that follow Thomas Aquinas and Hadewijch. Many Christian theologians will be far less confident in making clear the nature and inner life of God and therefore in making clear the nature of the soul's supreme union with God. But I agree with contemporary feminist critique that the same ideal and norm for good human life affirmed in this early and medieval Christian theological description will continue to dominate establishment Western thought, religious or secular, to the present day. The continuity is not only in theology and philosophy. It is the same norm for human life expressed in most lives of saints before the eleventh century. Almost all venerated saints were martyrs or virgins. It is the same norm expressed in existentialist fiction, in the heroes of *The Plague*, *The Fall*, *Nausea* and *Dirty Hands*. Modern atheistic philosophers and early and medieval theologians think the same way when they identify the supreme possible worth of an individual human life. They turn their

gaze as studiously as Thomas Aquinas from what women of their time
point to. I agree with Hegel that philosophy, theological or secular,
expresses much of the whole culture, for philosophy is "the mind of
the time grasped in concepts." But that is another story—though it
points to possible further significance of the present book.

What then is this norm for human life affirmed by traditional
Christian theologians by their description of supreme union with God?
If the theologians are right about this union, what characterizes the
best human life? For them the best human life is not interactivity or any
mutual affecting. *It is self-sufficiency through total dependence on the Self-
Sufficient.* The bliss, fulfilment, achievement, worth of the individual is
to become as self-sufficient as he or she can. Who is more self-sufficient
than the believing, faithful martyr or virgin? Or for that matter, the
humorous, serene Socrates going to his death in obedience to divine
call? Or Mother Teresa giving herself completely to the dying poor of
Calcutta?

First question: How does his dependence on God render a person relatively self-sufficient? The human individual becomes self-sufficient by holding fast to the absolutely self-sufficient divine. By facing
his or her total intrinsic insufficiency of self, the person reaches out
and shares the divine self-sufficiency. Given this intimate, complete
dependence on God, an individual is from every other point of view
wholly self-sufficient. The clinging may be intellectual contemplation
or vision, as Plato or Thomas Aquinas identifies it. It may be burning
love, as Bernard or Bonaventure identifies it. It may be complete merging in awareness, as Pseudo-Dionysius or Bernard, identifies it.

> As God willed that everything should be for himself [*propter seip-
> sum*], so, too, should we will that neither ourselves nor anything
> else have been or be except for him, i.e., because of his will alone,
> not our pleasure. There will give delight not so much either our
> calmed need or alloted happiness as that his will in us and for us
> is seen to be fulfilled....O love holy and chaste! O sweet and
> pleasant affect! O pure and serene intention of the will that is
> surely so much more serene and that in it no mixture of one's
> own is still left. So much the more pleasant and sweet as all that
> is felt is divine. To be so affected [*affici*] is to become [*deificari*]
> divine. [*As the water in wine, the redhot iron, the illumined air*]... so
> then in holy persons every human affection will have to in some
> ineffable manner liquefy from itself and be completely poured
> across into the will of God. Otherwise how will God be all in all if
> in the human person anything of the human person remain. The

substance will indeed remain, but in another form, another glory, and another power. (SBO III, *De Diligendo Deo*, 28:143).

Whether vision, love or merging, this is a secure connection with the divine. It is always also *mimesis theo*, "imitation of God." This, too, grounds invincibly the individual's self-sufficiency. The goal of all human life, the goal that encompasses all human values, indeed, the only good human life, is to be both as completely like the Self-Sufficient and as completely connected with It by knowing and loving as humanly possible. For Christian morality and spirituality, the self-sufficiency of God is key not only because it moors the self-sufficiency of the human individual. The self-sufficiency of God is also a model for the consequent human self-sufficiency. If God be not wholly self-sufficient, then the foundation and model of good human life collapses. The self-sufficiency of the human individual would collapse with it. Hadewijch's idea of her affecting God is profoundly threatening. It is incomprehensible, perverse, or obscene within Christian thought. The supreme model of human life includes vulnerability, divine as well as human!

Hadewijch's threat is yet broader. Her account of supreme fulfilment calls in question not only the tradition's account of human good but also the traditional account of the nature of *all activity and causality, human as well as divine*. I again bring together here, proposing for more coherent, longer meditation what I have already argued several times. The traditional concept of God's complete self-sufficiency is bound with a specific understanding of action and causality. We can consult here the *Summa Theologiae* of Thomas Aquinas, a synthesis of what in the thirteenth century was, and later continued to be, traditional Christian thinking. Thomas uses *agere, actio, actus*, and *activitas* in various senses, but the basic picture is clear. God in creating, saving, and so on is perfectly, completely active. Creatures are active to the degree in which their activity resembles God's. Humans are the most active beings on earth because their activity most resembles God's activity. But what is activity?

A being is active to the degree to which it self-sufficiently acts, i.e., *gives of what it has*. It is more active to the extent that it *gives without losing what it gives*. Activity par excellence is overflow. God's creating is paradigmatic activity for he gives being to all else while he loses nothing at all. The activity of inorganic material, e.g., the moving ball hitting and moving something or the hot iron heating something else, is the least activity because it loses all that it gives. It gives motion and heat but loses it in giving.

Moreover, to the degree to which one is active, ergo, self-suffi-

ciently active, overflowing, one *gives without gaining anything*. Giving
is not getting. In the higher activity the attraction or final causality of
the agent plays a greater role. The agent is also, or contains also, the
final cause. One acts on another by bringing him to imitate oneself.
Models deeply influencing Western thinkers from the beginning were
activities such as teaching, sculpting, and ruling. The teacher becomes
neither more nor less wise when the pupil learns wisdom from him.
The sculptor neither gains nor loses the beauty in his mind when he
carves the marble. The ruler becomes neither more nor less just by
making his laws and decrees just. The ideal is to be as active as possible and as little passive as possible, which means being as little recep-
tive as possible.

Modern atheistic versions are similar. Rieux gains and loses
nothing in serving the victims of *The Plague*. Roquentin will overcome
his *Nausea* and be "saved" if he writes and publishes a novel and
someone some day reads and appreciates it and him. But Roquentin
need never know this individual appreciation. Basically, the medieval
theologians according to whom God creates the world in order that
through it people appreciate God are the same. God does know
whether any actually appreciate his deed, but the actual appreciation
in no way affects God. So too, Socrates gives rich wisdom and inspira-
tion to his disciples and all people but gains and loses nothing.

The counterthought to the established medieval theology, counterthought exemplified by Bernard and other twelfth-century theologians
and, much more so, by Hadewijch, the affirmation of mutuality in
knowing and loving as the main human value, faltered and fell after
the high Middle Ages though some of it kept going and, I believe, is
being revived today. Whether the movement was crushed by champions of self-sufficiency or subsided of its own dynamics is beyond my
competence even to conjecture. But in Western thought, whether theological or secular, the glorification of self-sufficient individual life continued after the thirteenth century and even grew.

Martin Luther rejects scornfully the Greek structure of Catholic
theology. But he takes a further step away from mutuality and toward
greater individual self-sufficiency through utter dependence. He
expounds how thoroughly passive before God is the believer in all the
believer's real knowing and loving. Luther's act of saving faith is a
pure receiving of saving Love, a simple belief that God loves him. The
believer does not even have a choice in the matter. Persons believe
because God makes them believe. Free will, like reason, plays no part
in the connection with the Self-Sufficient. The Christian's love, then,
flowing irresistibly from the faith and passing on God's love is exclusively the Saviour's work and pure bounty.

But consequently, inasmuch as one truly believes, one is, in relation to other people, supremely active, purely giving. In giving out of the divine love received, the believer gains and loses nothing. The good tree bears good fruit, losing and gaining nothing thereby. The most important believer is the preacher, passing on God's Word, for faith in that Word meets all the believer's needs as the Word raises him to heights of faith and love. Luther's believer does nothing for God, does nothing to God, and certainly does not content God or sate God's hunger by believing in God's love nor by loving God with that love.[1]

For other modern Western thinkers, the dependence on God will be vague or fade into the background or completely away. Reason and/or free will often play a central part in the individual's sufficiency. But the model for good human life is still the self-sufficient individual, self-sufficient in his or her knowing and loving and in consequent giving to others. Consider the foundational work of Descartes or the ideal of Nietzsche. In the last two centuries, at the center of various Kantian moralities, religious or secular, stands the individual legislating to himself by an act of what Kant called both practical reason and will. The individual autonomously commands himself or herself to treat all humans as ends in themselves, independently of his or her getting or losing anything. The existentialist resembles the Stoic in putting the core of good human life in the individual's free, constructive will, a free will for all other free wills. Recall again the heroes of *The Plague* and *Nausea*. The two steps, or two moments of the first step, remain basically the same. First, the individual achieves autonomy. Second, in the autonomy, the individual wills good simply for others—in other words, self-sufficiency overflowing in pure giving. Recall, too, the café-keeper in the movie *Casablanca* and the sheriff of *High Noon*. The giving may cause great pain and risk great loss. But pain and loss are extrinsic to the self-contented will of the hero, willing the good of others. In this respect, popular twentieth-century stories differ little from the old *acta* of martyrs'.

For twentieth-century Christian theologians, Catholic or Protestant (e.g., Bonhoeffer and Karl Rahner), the supreme reality is divine, outgoing, compassionate love with which the individual fuses securely by his or her faith and love. As God so the individual Christian is wholly secure and wholly "for others." As the divine love so the human love is wholly for the other, not for self. As God gets nothing out of such loving except the loving itself so, too, the Christian. Thus, neither God nor the Christian seeks anything for self. God and Christian in their love have nothing either to lose or to gain. No one can give them anything or deprive them of anything. No one can content or satisfy them, for they have All, are contented, and satisfied by their outgoing love.

The intellectual struggle with which Bernard and twelfth-century theologians influenced by him were never finished came about because they wanted to be guided by analogy with something different from teaching, sculpting, and ruling. They were convinced that the courting and lovemaking by human lovers made an excellent image of the love and lovemaking of God and the soul. To what extent were they influenced by songs of courtly love and to what extent were those songs and their writing influenced by a new evolving experience of the time and culture? Perhaps, again, theology strove to grasp in concepts the mind of the time.

A giant standing on the shoulders of dwarfs, Hadewijch rose on the thought of preceding theologians and saw beyond their intellectual struggle. She thought out a unified, consistent kind of causality that is verified in erotic experience. I use "erotic" in the sense of Audre Lorde, Carter Heyward, Mary Hunt, and other feminists, that is, as a range of human embodied experiences which include far more than the sexual but for which the sexual is an apt paradigm. This is also the sense of Hadewijch's *Minne.*

For a medieval thinker, as for a modern thinker, it is not an easy question: how could a typically bodily relating like the sexual be an apt image of the most personal human happening? This lowest of bodily relatings, as I will discuss in Part II and the Appendix. How could it image their most intimate, supreme relating with God, the best possible happening in human life? Neither Bernard nor Hadewijch tackled the problem on the bodily side. They did not try to think out what then the bodily and the sexual must be. They tried to think out what then the personal must be.

It is hard, generally speaking, to know when a medieval thinker consciously endeavors to move beyond the tradition and lead it further on. The intellectual fashion in medieval times is to invoke authority and precedent for all novelty, as Thomas Aquinas brilliantly exemplifies. In this respect, we have to only wonder about the intentions of Bernard and Hadewijch. Consciously or unconsciously, Bernard in the early twelfth century led bands of the tradition on in his articulating the growing sense of Christians that the best of human life lay, not in the summit of something like intellectual contemplation, but in ecstasy of something like embodied, mutual love.

Hadewijch, as I keep arguing, went consciously or unconsciously beyond Bernard. She refused to reduce the best in life, divine or human, to self-sufficiency overflowing in pure giving to others. She stayed true not only to the Song of Songs and songs of courtly love but also to her own experience of love. God, being Love, must need others. Divine Love must include its desire and hunger for others, which only others could content and satisfy. It was a fact of experience: the best love, supreme love, was not completely satisfied self-love giving selflessly to others.

In so doing, Hadewijch articulated an ideal of sufficiency in mutuality. Hadewijch had complete confidence that the best in life, both human and divine, was indeed sufficient. But it was not a sufficiency of a self. It was a sufficiency of selves sufficing for themselves *and* each other. Preceding chapters have documented how for Hadewijch this sufficiency consisted in a mutual affecting of each self by the other. In the next chapter, we hear more of this affecting of each other. The human lover, in particular, is "proud" of this "deed" and the "strength" she exercised in it. Hadewijch frequently refers to this as her conquering of divine Love, while at the same time, Love conquers Hadewijch. What she means by this is not only a remarkable event in the history of Western thought but also a challenge to late twentieth-century thought. She challenges thinkers carrying on the millennial intellectual tradition and thinkers opposing some of that tradition, e.g., its feminist critics. As I submit in the following chapter, Hadewijch brings out a side of the best loving mutuality that is only slightly developed in the contemporary feminist literature I have read.

# 5

Hadewijch, while often echoing twelfth-century theologians of love, speaks at times as neither they nor earlier theologians spoke. How do God and the soul make their final step to supreme embrace? Hadewijch echoes Bernard in affirming repeatedly that we reach final union with God by first loving God as human being, as Jesus Christ. "God granted me to know the perfect pride of love; to know how we shall love the Humanity in order to come to the Divinity, and rightly know it in one single Nature. This is the noblest life that can be lived in the kingdom of God" (Vision 12, 188f.). But Hadewijch uses a phrase here that Bernard does not.

Unusual for early Christian theologians is Hadewijch's phrase "the perfect pride of love." The usage of "pride" in a positive sense is almost unknown in earlier Christian theologies. Of the Church Fathers, only Jerome speaks of good pride. He does so in two texts that Hart judges may be the source for Hadewijch's usage. But Hadewijch uses the noun *fierheit* with the meaning of pride in the good sense nine times and far more frequently the corresponding adjective *fier*. Such speech was also influenced undoubtedly by courtly literature, as was similar language, by other devout women of the time, such as Beatrice of Nazareth.[1] To my knowledge, what Hadewijch means by the phrase is, as we shall see, not found in earlier Christian theologies at all. Hadewijch affirms similarly elsewhere a good "pride" in the strongest human love of God (e.g., Vision 10, 54ff.; Vision 12, 152ff.; Vision 14, 172ff.; Poem in Couplets 1, 63–94). What does she mean?

Hadewijch is proud of the strength of her love for Christ, God, and man. At the conclusion of Hadewijch's Vision 14, God, in a voice of thunder that silences all else, addresses her:

O strongest of all warriors! You have conquered everything and opened the closed totality, which never was opened by creatures who did not know, with painfully won and distressed Love, how I am God and Man! O heroine, since you are so heroic, and since you never yield, you are called the greatest heroine! It is right, therefore, that you should know me perfectly! (Vision 14, 172ff.)

By her strong, unyielding love Hadewijch has "conquered everything." The perfect knowledge, which is now right for her to have in supreme union with God or in fruition of God, is described earlier in the account of this vision as well as in the account of other visions she has had (e.g., Vision 14, 145ff. and Vision 12, 152ff.). Because Hadewijch is so heroic and never yields, what is right for her and not for most others who love God is that she have this fruition during her earthly life.

Unitive fruition of God is the exception in Hadewijch's life. Her love sustains her in a life of terrible suffering. In her love, she suffers, above all, from privation of the fruition, but she continues to love. It is the strength and power she has to love that wins her occasional fruition of God as it also profits others.

Behold, this is my bride, who has passed through all your honors with perfect love [*volmaecter minnen*], and whose love [*minne*] is so strong that, through it, all attain growth!...Behold, Bride and Mother, you like no other have been able to live me as God and Man....

...you like no other have superhumanly suffered much among men. You shall suffer everything to the end with what I am, and we shall remain one. Now enjoy fruition of me, what I am, with the strength of your victory. (Vision 10, 54ff.; cf. Vision 3, 1ff., and Vision 1, 265ff. and 288ff.)

Recall what we learned in an earlier chapter of Hadewijch's understanding of Love. Exercising "painfully won and distressed love," passing Love on to others, serving others, being thus like Christ in his earthly life, is essential. It is part of loving, of being possessed by Love (Letter 6, passim; Vision 1, 138ff.). It is part of what Love loves, wants, and enjoys. But it all aims at one all-dominant goal for Hadewijch as for all humans: to be united with Love in fruition (e.g., Letter 1 passim). In Vision 14, a few lines earlier than the lines quoted above, Hadewijch knows of Christ, her Beloved, "how in fruition he embraces himself." She has been able "to taste Man and God in one knowledge, what no man could do unless he were as God, and wholly such as he was who is our Love."

This is what Hadewijch expects to have for all eternity. This is what she can possess temporarily in earthly life (Vision 5, 59ff.; Vision

6, 92ff.). It is also to be at the same time what she is and what God is! (Vision 7, 1ff.; Letter 7, 1ff. ) This understanding of the single, all-satisfying goal for all humans is no historical innovation. Bernard of Clairvaux offers much the same picture. Bernard's emphatic assertion of how completely the human loves *become* Divine Love in supreme fruition compels Etienne Gilson to spend much of his book (1940) refuting charges that Bernard becomes pantheistic and denies human individuality here.

However, that God affirms and praises, and Hadewijch is proud of, her all-conquering strength and power of love is not to be found in earlier theologians. It fits well with what we have already explored to some extent: a mutuality in which God and Hadewijch affect each other and give to each other. This all-conquering strength and power of Hadewijch is a further dimension of the mutuality studied so far. Let us look at the conquest by which Hadewijch won the right to and enters upon this final fruition. In what exactly was she so strong and powerful that she gained fruition?

The overarching goal of Hadewijch, her fruition of God, is also God's goal for her. It is God who makes the goal possible for humans. Only the obedient, serving life and love of the Son of God made Man can perfect what was wanting "on our part." He uplifted us, drew us up by his divine power and his human justice to our first dignity and our liberty. In this liberty we were created and loved. To it we are now called and chosen in his predestination, in which he had foreseen us from all eternity (Letter 6, 324ff.). Hadewijch's strength of victory, celebrated, as quoted above, in her accounts of Visions 10 and 14, is declared at the beginning of the latter account to be a "new strength," "a new state of power," infused into Hadewijch by God. Her strength was, indeed, "the strength of his own Being, to be God with my sufferings according to his example and in union with him, as he was for me when he lived for me as Man. That was the strength to endure, as long as the fruition of Love was denied me" (Vision 14, 1ff.).

But Hadewijch's progress proceeds also according to her own energy and power. Her progress to her goal is *her* doing, too. She achieves it by her free choice. She contrasts herself with the "saints" who, enjoying fully and definitively the fruition of God, will necessarily (not freely) to have that fruition. Hadewijch freely wills it.

> For I am a free human creature...and I can desire freely with my will, and I can will (*willen*) as highly as I wish (*wille*), and seize and receive from God all that he is, without objection or anger on his part—what no saint can do. (Vision, 11, 98ff.)

When I could thus turn myself against him [God], by wishing to free all men in the twinkling of an eye, otherwise than in accordance with how God had chosen them, it was a beautiful and free expression of life as a human being. Then I could desire what I wished [*woude*]. But when I did the opposite, I was more beautiful and taken up into a fuller participation in the Divine Nature. (Vision 11, 174ff.)

Similarly, in Vision 1, 288ff., Hart's "wish," as also Hart's first "wish" in the Vision 11 passage quoted here in the main text, translates *wil*, which means basically "will." Similarly, in Letter 2, 163ff.: "Thus you may become perfect [*volmaect*] and possess what is yours!—if you wish [*wildi*]." "If...you wish (*wildi*) to have what is yours" starts the final paragraph of the letter.

Hadewijch *can* exercise free will while the saints cannot for they "have their will perfectly according to their pleasure; and they can no longer will beyond what they have" (Vision 11, 98ff.). Hadewijch has to exercise her free will if she is to will all that God is, for to so will she must will to increase that desire in her, the desire that is indeed both divine Love and agony.

But in striving for this I have never experienced Love in any sort of way as repose; on the contrary, I found Love a heavy burden and disgrace. For I was a human creature, and Love is terrible and implacable, devouring and burning without regard for everything. The soul is contained in one little rivulet; her depth is quickly filled up; her dikes quickly burst. Thus with rapidity the Godhead has engulfed human nature wholly in itself. (Vision 11, 121ff.; cf., e.g., Vision 7, 1ff.)

Indeed, when God constantly invites individuals to unity in the fruition of himself, some are frightened by this fearful invitation and just warning and go away. Others, the proud souls "stand up with a violent new will" and raise themselves toward him (Letter 22, 39ff.). At times Hadewijch implies that by her power of free will she gains yet more such power, a direct share in God's universally ruling will:

And then the Angel said again to me: "O powerful and strong one, you have conquered the powerful and strong God, from the origin of his Being, which was without beginning; and with him you shall wield power over eternity in eternity! Read and understand!" And I read and understood. On each leaf was written: "I am the power of the perfect will; nothing can escape me." (Vision

1, 60ff. Cf. Vision 8, 98ff.; Vision 10, 54ff.; Vision 12, 152ff., though
none of these passages speaks of Hadewijch's conquest over God.)

In any case, that only Hadewijch's free will maintains and makes
potent her desire is one facet of the personal agency and power
Hadewijch experiences in her mutuality with God.

That to be fully possessed by Love Hadewijch can and must
choose freely is no theological innovation. Bernard of Clairvaux made
free will the center of a person's being, or "image of God" (SBO III, *De
Gratia et Libero Arbitrio*, 6, 19, 180). Recall Christian admiration, lively
in medieval times, of the lives of saints. Their chroniclers and other
admirers did not hesitate to speak of the saints' "victory" in achieving
personal sanctity on earth and bliss with God after death.

Some Islamic theologians maintained that creatures never act,
much less will freely, for that would derogate from the power of God in
the world. All activity of creatures is really only of God. When creatures seem to act, it is God alone who acts on the occasion of the creatures and their created situations. In refuting these "occasionalists,"
Thomas Aquinas argued that the more independently of God a creature acts, the greater the power of God is manifested. How powerful he
must be to ground and enable such relative independence! How great
is the achievement by God and thus the more glory the creature gives
to God. Free will is God's greatest creation, for by it the creature
achieves the greatest independence in dependence on God. God can
and does enable the creature even to disobey God. Thomas does not
innovate in any of this, but, vis-à-vis the occasionalists, articulates inci-
sively this central Christian theological tradition (ST I, 105, 5; ST I, 115;
SCG III, 69).

That an individual is praised for an achievement by her strong
will is, therefore, theologically traditional and commonplace in
medieval Christian piety. Not so traditional or commonplace, I submit,
is, first of all, that Hadewijch herself reports and accepts this praise of
herself (e.g., as "heroine") and her achievement. The fact that she does
so and that the quotation immediately above from Vision 1 is the entire
speech of God concluding the vision is untraditional in its dramatic
emphasis on Hadewijch's power and agency. One might dismiss the
praise of herself as the self-centered vagaries of an eccentric, frustrated
woman. Or one might, as I do, wonder whether we hear here a woman
who, out of her genuine, willed, achieved experience and against the
ambivalence and taboos of the tradition, affirms confidently something of her real power and agency as a person vis-à-vis another person, indeed, a divine person.

But what of the "conquest," which God in this text says Hadewijch has achieved? What kind of conquest is it by which she gains final access to perfect knowledge of and supreme union with God? In other passages Hadewijch fills in the picture. The picture is in form and content unusual—in my reading, unprecedented—in Christian tradition. Hadewijch's conquest that opens the entirety of God for her (in those hours or days when she reaches and has her "visions") was achieved by her through unremitting battle with divine love. All who love God must wage this battle through their entire life. "Love alone is the thing that can satisfy us, and nothing else; we must continually dare to fight her in new assaults with all our strength, all our knowledge, all our wealth, all our love—all these alike. This is how to behave with the Beloved" (Letter 7, 4).

The Beloved is Christ, who is God, or Love. It is Love itself that Hadewijch and other successful warriors fight and conquer: With what do they fight Love? With their love. "But he who dares to fight Love with love [*minne met minnen*] / Very quickly comes to his goal" (Stanzaic Poem 38, 7–8. In other words: human beings conquer God by their ardor and passion. "If anyone dare to fight Love with ardor / Love cannot resist the violence of the assault" (Stanzaic Poem 38, 58–59; cf. 39, 91–95). So, too:

> ...the noble soul in Love's service lives so free
> That it dares to fight [Love] with passion
> To the death or nearly,
> Until it conquers the power of Love.
> (Stanzaic Poem 40, 21–24.)

Who conquers the power of Love is "a champion" (25–29). What has empowered the champion to conquer love? The champion's free choice and will, as seen above, and the very passion and ardor of the champion's love. It is by love, freely willed and violently passionate, that Hadewijch and others can conquer love. How does a human being conquer Love by love? What does it mean?

# 6

In insisting that all who love God must strive to conquer him, Hadewijch invokes in Letter 12 the Bible story of the patriarch Jacob wrestling with the angel (Gen. 32:24–31). In so applying the story of Jacob's fight with God, Hadewijch is inspired probably by earlier use of that story by medieval thinkers. Hadewijch may have been influenced by sermons such as the second sermon of Guerric of Igny for the feast of Saint John the Baptist (167–69). How Hadewijch's application of the story of Jacob's wrestling coincides with, and how it differs from, Guerric's are suggestive for our purpose. Further comparison of other interpretations of the Bible story before Hadewijch's time might well be enlightening.

Guerric's sermon is on the text of Matt. 11:12: "From the days of John the Baptist the kingdom of heaven suffers force, and the violent carry it away [*vim patitur et violenti dirimiunt illud*]." Guerric takes the patriarch Jacob to illustrate this salutary exercise of force and violence against the kingdom of God. Those who would gain the kingdom of heaven must imitate the patriarch who struggled with God through the night until morning. Jacob held God resolutely though God asked to be let go. "I will not let you go," said Jacob, "until you have blessed me."

Guerric's Bible text, common in the Middle Ages, has Jacob wrestling with an "angel." Guerric explains that Jacob wrestles really with God: God is in the angel. Otherwise, argues Guerric, Jacob would not have said later, "I saw the Lord face to face." For Guerric, it is also significant that when Jacob's divine adversary "saw that he could not conquer [Jacob], he touched the nerve of his thigh and he forthwith became lame."

Hadewijch's application of Jacob's wrestling with God is very similar to Guerric's. The wrestling illustrates that those who want to love God, as Hadewijch says, must fight and conquer God by love. Guerric notes, though only in passing, that those seeking the kingdom of heaven seek him with love, whereas for Hadewijch, the love of those seeking God is a constant refrain. For both Hadewijch and Guerric, the final laming of Jacob illustrates that lovers of God conquer God by love in order to be conquered by him. Some, therefore, of the inter-subjectivity and mutuality we trace out in Hadewijch is found already in and may come from predecessors like Guerric. Hadewijch and Guerric affirm, too, that God himself has ordained this fight with and victory over him as the way for human beings to achieve final, blessed union with God (Letter 12, 174ff.; Vision 13, 46ff.). In Vision 14, imme-diately before God's speech quoted above, Hadewijch acknowledges that she has reached the final fruition because she has been chosen.

But there are differences between Guerric's and Hadewijch's application of Jacob's wrestling with God to human lovers' conquest of God. The following traits of Guerric's description of human con-quest of the divine are absent from Hadewijch's. I follow closely here the wording of Guerric:

1.  For Guerric, the conquest by those who carry off the king-dom of heaven is violent because it takes by virtue what was not due to nature, even to the nature of the innocent. It is also violent for those who were by nature children of wrath and hell to force themselves, by penance, "inferior work" (*labore improbo*) that conquers all, into the inheritance of the saints and the sharing of glory. For Hadewijch, as we saw, it is the "violence" of love's ardor and passion that conquers.

2.  For Guerric, this salutary violence is characterized by penance aimed at the mortification of the body.

3.  This making dead of the body, we cannot achieve unless God has "approved for himself" (*probaverit sibi*) our invincible constancy combating Him. The combat is primarily a pre-arranged test of the human combatant. This imagery, influ-enced surely by contemporary chivalric ideals and practices, is present in Hadewijch's account, too, but it is not the pri-mary dynamic in her conquest of God.

4.  What does God's resistance to our attacks consist of? God's resistance is of two kinds for both Hadewijch and Guerric: (a) God withdraws from us when we want to draw closer; we cry out and he does not hear us; and (b) God also carries out positive harsh actions against us. Guerric's phrases are: God dirties us after we cleanse ourselves, gives us the oppo-site of what we want, opposes us with a rough hand, fights back against us. A principal difference here between Hadewijch and Guerric is that at the last, loftiest stage of her

assault, God's resistance is described by Hadewijch *only* as staying away. As we will see, Hadewijch thus puts in relief what she sees to be God's motive for finally yielding to the human lover. It is not that God judges that the individual has passed a test.

5. For Guerric, all this counterattacking (as above in no. 3) by God is "dissembling mercy" (*dissimulatrix clementia*). God's hardness against those seeking the kingdom is pretense (*quae duritiam te simulas*). It is actually in "fidelity" (*pietate*) to the seekers that God fights against them. God hides in his heart the love he has for those who love him. Great is the quantity of sweetness he conceals for those who fear him. As we will see, in Hadewijch's description of her and others' conquest of Divine Love, she does not assert or imply that Love dissembles or that God conceals the extent of his Love for the individual. On the contrary, her final conquest of Divine Love brings about a real change in God's Love for her.

6. Knowledge of this "hidden" attitude of God, continues Guerric, should fortify those seeking the kingdom. Those who love and fear God should not despair at his harsh treatment of them, but trust that he still loves them. God loves to suffer force from them. He loves to let them conquer him. Even when, as in the Bible, God is angry and extends his hand to strike, he looks for a man, like Moses, who will "stand up to him" (*resistat sibi*). If he does not find one, he complains or laments. He wants his human adversary to be stronger and prevail. Hadewijch says something similar, for instance, in the conclusion of Vision 14 quoted above or when she encourages "humility" and "reason," as we discuss below. But at times she says something contrary. As we will see shortly, Hadewijch triumphs over God finally in *mistrusting* God's love.

7. To encourage the human lover/seeker to combat God all the more vigorously, Guerric affirms God's hidden love not only in a general way. Guerric spells out specifically what Divine Love aims to achieve through the combat, with what purpose God makes things hard for us in our very pursuit of the kingdom. God does so to sharpen our spirit to be strong and large, to exercise our "forces" (*vires*), to test our constancy, to multiply our victories, to add to our crowns. God thus works to give us strength and confidence in our strength so that we

are frightened by no adversities. Again, we see that for Guerric this whole conflict with God is primarily an exercise that
God requires of the seeker of the kingdom. Guerric has perhaps in the back of his mind the way a knight trains appren-
tices to knighthood who were in his charge or a lady testing
her would-be knight.

Hadewijch's description of her conquest over God has, we saw,
similarities with Guerric's description of the conquest over God by
those who seek the kingdom. Love is crucial in the fight with God.
Hadewijch's love is Divine Love (*minne*) that God shares with her. It is
thereby her love, too. It is her strength and power.[1] By her strong, powerful love she conquers Love. She reaches supreme fruition through
her powerful, battling love. Though Guerric uses the word "love"
rarely in the sermon we read, he seems to imply roughly the same as
Hadewijch in this respect, for both Hadewijch and Guerric see the conquest wrought by the believer's desire to be one with God, and Guerric refers to such seekers as "loving" (*diligentes*) (168C) and "armed by
the virtue of love" (*virtute dilectionis armatus*) (169A).

But not only does Hadewijch's account of the believer's conquest
over God lack, as we listed above, certain assertions by Guerric.
Hadewijch affirms what Guerric neither affirms nor implies. How,
according to Hadewijch, do she and others like her manage with their
love to conquer Love? For Guerric. it is by loving deeds: penance, mortification, virtuous action, and so on. For Hadewijch, on the contrary, it
is (A) by love (*minnen*) itself. But how?

As we noted above, it is the violence of the lover's ardor and passion ("nearly to death") that overcomes Divine Love. In yet another
phrasing: one fights and conquers Love *with longing*:

> And if anyone *then* dares to fight Love with longing [*niede*],
> Wholly without heart and without mind [*sinne*],
> And Love counters this longing with her longing;
> That is the force by which we conquer [*ghewinne*] Love.
>
> (Stanzaic Poem 38, 53–56; emphasis added)

> He who dares to fight Love with longing,
> Whatever cruelty he meets with,
> Shall possess her immensity.
>
> (Stanzaic Poem 39, 88–90)

These human beings conquer Divine Love by the force, the passion and ardor, of their longing. They dare to long for Love wholly

without heart or mind. This is their daring power that gains them victory over God. This is what Love "cannot resist." "If anyone dare to fight Love with ardor / Love cannot resist the violence of the assault (Stanzaic Poem 38, 58–59). The longing by the human lover for Love B: affects irresistibly Divine Love. The human longing C: makes Love counter with her own longing. It is a moment of intersubjectivity through which God and the soul move into their final union.

Guerric does make decisive in the battle between God and believer that God cannot resist "love" (*charitati*) (169A). But it is God's own love in himself that he cannot resist. In "battle" with the human lover it is simply to God's own Love that he yields, just as God's own love weakened his strength and led him conquered to the death of the Cross. Guerric does not say that God yields in the combat with human love because he is overcome or otherwise affected by the seekers' love, longing, ardor, passion, or whatever (= A above). Nor does Guerric say that God so yields because he responds, cannot but respond, to the seekers' longing with his own longing for them (= B and C above).

What gives Hadewijch's conquerors of God the strength to battle God with such bold longing? Is it, as for Guerric, a strength fed by confidence? For Guerric, the confidence comes from the realization that God's resistance is feigned. It is an act. The battle is only an exercise intended by God to test the seekers and to make them grow in strength. God "hides" his motives, but they know them. Hence, they can battle confidently.

Hadewijch speaks often in a similar vein. She encourages herself and others to have greater confidence in God's loving plan for them, but in the height of the battle with Love, as Hadewijch says several times, the lover's strength comes from the contrary of confidence. One factor that decides in the lover's favor the battle with God is her or his lack of confidence, or "mistrust" (*ontrouwe*) of God and his Love. Hart translates *ontrouwe* as "unfaith." Let us look at it more closely.

In Vision 13, Hadewijch learns how Love is conquered most completely. It is conquered by the love of a tiny elite, including Hadewijch, who had "many more wonderful deeds than all others had had" (Vision 13, 159f.). But this final conquest of theirs is not by deeds. They conquer Love by three voices of their love: their humility, reason, and "unfaith" (*ontrouwe*). The three voices, as Hadewijch describes them, contradict each other since reason is said to include faith and confidence in God, but together they constitute the power that conquers love. It is "unfaith" that interests us particularly. By not believing in Love, these individuals conquer Love! Hadewijch seems to me to identify unfaith as *the* decisive factor, but the text is perhaps not that unequivocal. Unfaith is certainly *a* decisive factor immediately preceding and bringing about Hadewijch's victory.

> And Mary...said to me: "Behold, everything is fulfilled! Penetrate all these attributes and fully taste Love. For you cherished Love with humility, you adorned and led Love with loyal [*ghetrouwer*] reason; and with this lofty fidelity [*trouwen*] and this entire power, you vanquished [*dwonghes*] Love and made Love one. Through this, and on account of your lofty power, is this secret heaven thus made known to you. Love, as you see her here, is thus adorned and praised by this song. For the denial of Love with humility is the highest voice of Love. The work of the highest fidelity [*trouwen*] of reason is the clearest and most euphonious [*bequaemste*] voice of Love. But the noise of the highest unfaith [*ontrowen*] is the most delightful [*suetste*] voice of Love; in this she can no longer keep herself at a distance and depart. (Vision 13, 214ff.)

Hadewijch in this passage does not make clear how humility, reason, and unfaith interplay in the elite lovers' conquest of Love. Clear is that each of the three contradicts the other two. They are therefore sequential, or at least alternating, phases of love. When humble, the lovers declare that they are not serving and loving Love. They know themselves to be nothing. They deny of themselves all love. Wholly annihilated in humility, they can nevermore believe they can attain Love's affection (Vision 13, 66ff., 124ff.).

The elite lovers cast off humility and place knowledge between themselves and God. It is knowledge of his power and kingdom, his goodness, his sweetness and his whole being. This knowledge replaces their humble denial in that it reassures them that God has empowered them with love and they can achieve Love. Thus Hadewijch in the passage quoted above refers to reason as "believing." I would translate thus *trouwen* and *ghetrouwer* as "belief" and "believing" (Hart in the quotation above uses "fidelity" and "loyal"), since Hadewijch intends here evidently a contrast with the immediately following "unfaith" (*ontrowen*) (159ff.).

In Stanzaic Poem 4 Hadewijch depicts this same faith of reason, but here it is the whole picture. She is presumably describing the lot not of the elite, but of the majority of lovers of God. It would seem to be also Hadewijch's own experience before or in abstraction from the visions in which she attains supreme fruition of God. It would seem also to be the experience she at times echoes of twelfth-century theologians such as Bernard of Clairvaux and Richard of Saint Victor.

In this poem, Hadewijch extols the faith that reason gives. Hart again often translates *trouw* as "fidelity" throughout the poem. The translation is defensible: Hadewijch urges fidelity and loyalty to Love. At the same time, however, every one of her nine stanzas describes or refers to the suffering of the longing lovers and encourages the longing lover to have confidence and rejoice in promised bliss after death. "Faith" or "trust" would translate *trouw* at least as well. E.g.

> O hearts, let not your many griefs
> Distress you! You shall soon blossom;
> You shall row through all storms,
> Until you come to that luxuriant land
> Where Beloved and loved one shall wholly flow through
>     each other:
> Of that, noble fidelity [*trouwe*] is your pledge here on earth.
>
>                     (Stanzaic Poem 4, 43–48. Cf.
>             Stanzaic Poems 35 and 36 and Letter 6, 350ff.)

To work by *trouwe*'s counsel is to conform oneself with truth, so one "is accompanied by works of truth." In the final stanza of Stanzaic Poem 4 Hadewijch sums up the poem by assuring us that

> God must give noble souls an insight
> That will enlighten for them the life of exile,
> Since they are now wounded and driven from their goal...
>
>                                         (49–51)

Hart translates by "insight" *redennen*. This is the same word translated as "reason" in the passages above from Vision 13. In Letter 2, similarly, Hadewijch encourages the fear that we do not serve Love sufficiently, but faults any accusation that we have loved enough and that it is Love that is "unfaithful" (*ontrouwen*), helping and loving us too little.[2]

But as we saw in Vision 13, the elite lover at times abandons humility and believing reason for unfaith. In Letter 1, Hadewijch describes this unfaith without indicating its positive effect.

> Alas, dear child! although I speak of excessive sweetness, it is in truth a thing I know nothing of, except in the wish of my heart— that suffering has become sweet to me for the sake of his love. But he has been more cruel to me than any devil was. For devils could not stop me from loving God or loving anyone he charged me to help forward, but this he himself has snatched from me.

What he is, he lives by, in his sweet self-enjoyment, and lets me thus wander far from this fruition, beneath the constant weight of nonfruition of Love, and in the darkness where I am destitute of all the joys of fruition that should have been my part.

Oh, how I am impoverished! Even when he had offered and given me a pledge of the fruition of veritable love, he has now withdrawn—as, in part, you well know....

...now my life is like his to whom something is offered in jest, and when he wishes to take it his hand is slapped, and he is told: "God's wrath on him who fancied it true!" And what he supposed he held is snatched from him. (56ff.)

As Hadewijch says in the tenth Poem in Couplets, unfaith, unlike humility or reason, refuses to accept God's withdrawal and staying at a distance. After explaining from several angles why lovers should accept suffering and unsatisfied desire, Hadewijch continues to the contrary:

> Desires of love, moreover, cannot
> By all these explanations be quieted.
> Desire strives in all things for more than it possesses:
> Love does not allow it to have any rest:
> Even if all the suffering were massed together
> That ever was, or is, or shall be,
> It could not conquer so much
> As desire of veritable Love can.
> Desire...undergoes pressure from noble unfaith,
> Which is stronger and higher than fidelity [*trouwen*];
> Fidelity, which one can record by reason,
> And express with the mind,
> Often lets desire be satisfied—
> What unfaith can never put up with;
> Fidelity must often be absent
> So that unfaith can conquer;
> Noble unfaith cannot rest
> So long as it does not conquer to the hilt;
> It wishes to conquer all that Love is:
> For that reason it cannot remain out of her reach.
> Consequently the soul feels much bitterness,
> Which Love could heal in a short time.
>
>                                        (75–98)

The pieces of the picture fit together. Unfaith arises out of frustrated desire. In the fury of unsatisfied desire the lover will have nothing of humility and trusting reason and their resigned acceptance. Flaming desire turns then to bitterness against Love as well as to unfaith in Love's pledges Unfaith, however, is still a voice of love because it spurs on, or indeed is, love's desire for Love.

In Vision 13, the lover's voice denies its love humbly, sings of Love with faithful, trusting clarity of reason, and doubts Love sweetly. In humility, the elite lovers disavow that they love God. By reason, they place their knowledge as knowing, believing resignation between themselves and God. By unfaith, they call constantly for unitive fruition, for they "did not believe in the love of their Beloved; it rather appeared to them that they alone were loving [*minden*] and that Love [*minne*] did not help them. Unfaith made them so deep that they wholly engulfed Love and dared to fight her with sweet and bitter" (179f.).

Bitter is their discontent with what they had. D: Sweet is that Love's depriving them enriches them with great strength to follow love's demand that they be as great as she. Sweet, therefore, is their untrusting and therefore yet stronger desire for fruition of Love. E: This sweetness of desire presumably makes the sweetness of the voice of unfaith from which Love, as we read later on in the Vision (quoted above) and in the tenth Poem in Couplets, cannot stay away from or depart. It is an important question, whose answer lies outside my present knowledge, to what extent Hadewijch innovates theologically in making the conclusive free choice for full union with God to be a choice not against the attraction of created goods but against the security of accepting the agony of unsatisfied love of God. The question is all the more suggestive in that Hadewijch's desire to step up the desire even more does finally get for her, as detailed below, wonderful satisfactions of the desire, satisfactions that make the agony worthwhile. I say "important" and "suggestive" in light of the overall problematic of the book: what new image of a good human life do Hadewijch and other contemporaries press toward the center of intellectual and cultural attention?

Hadewijch's account of her conquest of God differs from anything I know of in Christian writing up to or during the Middle Ages. I have not found the like in, for instance, the writings of Guerric, Bernard, or Richard of Saint Victor nor in Hildegard of Bingen or Beatrice of Nazareth, who probably lived a generation later than Hadewijch. Decisive in Hadewijch's triumph over Love is her lack of faith, her refusal to trust in God's Love. More accurately, it is not her unfaith as such that triumphs over Love. It is her unfaith *qua* rising from and in desperation increasing her desire or longing for God. It is her resultant desire/longing that engulfs Love, and Love cannot stay away from such sweet, mighty love.

This dynamic of unfaith is affirmed also in Letter 8. By unfaith, "we fear that Love does not love us enough because she binds us so painfully that we think Love continually oppresses us and helps us little, and that all the Love is on our side" (27ff.). This "unfaith" (*ontrouwe*) lacks the "faith" (*trouwe*, which Hart translates here as "fidelity") that allows us to rest peacefully without the full possession of Love or to take pleasure in what we have in hand. Thus, as in Vision 13, F: unfaith swells our desire for God.

> This noble unfaith greatly enlarges consciousness, Even though anyone loves so violently that he fears he will lose his mind, and his heart feels oppression, and his veins continually stretch and rupture, and his soul melts—even if anyone loves God so violently, nevertheless this noble unfaith can neither feel nor trust [*trouwen*] Love, so much does unfaith enlarge desire. And unfaith never allows desire any rest in any fidelity [*trouwe*] but, in the fear of not being loved enough, continually distrusts (*mestrout*) desire. So high is unfaith that it continually fears either that it does not love enough, or that it is not enough loved. (Letter 8, 27ff.)

What then is the outline we have traced so far of that moment of intersubjectivity in which the human person conquers God:

A: Human lovers, though their love is a share in Divine Love, can conquer Divine Love by the ardent, violent longing of their love.

B: Human lovers thus affect Divine Love.

C: So that Love responds with its longing for the lovers.

D: Human love affects Love here by the human lover's unfaith in Divine Love (the lover doubts that Love loves the lover and that the lover loves Love enough).

E: For irresistibly sweet to Divine Love is the stronger, ever more violent desire of the human lover for Love.

F: This desire both generates the lover's unfaith and becomes stronger, more violent, because of this unfaith.

How does the lover's unfaith move Love to respond? It is tempting to interpret Hadewijch as implying that the voice of unfaith affects

Divine Love because it expresses shame or blame ("You don't love me!) or challenge ("Show me!") or appeal for pity ("Look at poor me desperate for you!). I find no evidence for such interpretations. We have to hear what Hadewijch says: Divine Love is moved by, cannot resist, the sweet violent longing of the lover to be with Love. This is why unfaith also doubts its own love; it believes that if it desired enough, Love could not stay away. This is what makes the voice of unfaith sweetly overpowering to Divine Love: it expresses the extreme longing of the human being to be with Love.

Hadewijch cries to Love, "I don't believe you! You don't love me!" Love hears a desperation, that rides a fiercer and fiercer longing for Love. "I want you so much!" Hadewijch's desire is so violent that it has left behind humility and reasonable faith and burst into itself alone, pure desire and intensity of passion. The Other cannot resist it.

On the other hand, G: this is but one moment in the supreme intersubjectivity. Hadewijch conquers Love only because she has been conquered by Love and only in order to be, again and far more completely, conquered by Love.

The intersubjective dynamic to which Hadewijch testifies is, of course, a common human drama. "Every day, everywhere, all around us, in films, soap operas, popular songs, etc., as in stories of courtly love in her time. Oh, yes, in real life, too. Particularly with women. Banal. Well, moving, too, in a way. But not really important. Sort of melodramatic, sentimental, emotional, subjective. Not part of what makes human life really worthwhile, satisfying, of dignity and meaning."

Hmmm. Let us listen again and try to pick up more exactly what characterizes this "banal" interchange between loves, this interchange which Hadewijch judges to be the most worthwhile, worthy, meaningful, satisfying reality in human life.

# 7

Hadewijch's successful conquest by her unbelieving love over Divine Love leads to a curious conclusion. In the main, unfaith had the truth. Humility and reason were, at least in large part, wrong! Both sides of the *ontrouwe* were right: Love *could* love Hadewijch more than Love did, for Love, now conquered, does love her more. Hadewijch *was not* loving Love enough because she did then make herself love more, i.e., desired more to possess completely Divine Love. This new stronger longing gets that possession. It is not the unfaith, but the increased desire/love, exploding into unfaith and then spurred on by unfaith, that effects the final conquering. Yet the conquest proves the unfaith was true in the first place. And Hadewijch's traditional Christian view of human life was in error!

What then is the new view expressed in her conquest? Let us take Hadewijch at her word. Let us take her words not to be flailing verbiage or uncontrolled rhetoric. Suppose she meant what she said. What then is her experience, which she works to convey to the reader?

(a) In Hadewijch's unfaith (which she prizes and praises), *she leaves behind, like trapezes, the reassurances of solid reason and worthy faith/trust.* She does not leave them behind either definitively or provisionally—nor by any reasoning or by a new faith. She simply lets the reassurances slip out of her mind or lets them slip to the edge of her consciousness where they no longer affect the rest of what she consciously knows, feels, and does. This is all the more "irrational" because she dismisses conscious knowledge. Hadewijch does not gain or exercise the "eyes of love," made much of by traditional thought, because her increased desire enables her to see nothing new except how much Divine Love is desired by her. For the intellectual tradition, Hadewijch acts like a brute animal!

Lovers like Hadewijch reject the most basic reassurances, "for they did not believe in the love of their Beloved; it rather appeared to them that they alone were loving [*minden*] and that Love [*minne*] did not help them" (Letter 13, 179ff.). But this is an error. Hadewijch

65

asserts often elsewhere that God loves her and helps her. Hadewijch's desire that generates the unfaith has the characteristic of bodily passion: the impetus to exclude prior knowledge from consciousness and to induce error. No spiritual passion would do this.

(b) Why does Hadewijch throw aside the reassurances? Because of what she fills her consciousness with. Hadewijch *fills her consciousness at these moments of unfaith with the present moment*. Why? Because both her rising desire and her willed choice is to live wholly in this present moment of time and get now what she wants. Hadewijch is only peripherally, ineffectually conscious, if conscious at all, that she is sure of bliss with Divine Love in everlasting life after death. She is mainly conscious of *what is going on between herself and Love here and now*. In Part II, we will note that such immersion through overwhelming feeling in a present place and moment characterizes much medieval women's piety, exemplified by prayer before the *Pieta* or union with Christ on receiving the Eucharist.

(c) As with the reassurances of God's present hidden Love and Hadewijch's eternal bliss in the distant future, so too, Hadewijch is *only peripherally and ineffectually conscious of what love, satisfaction, and pleasure are at hand at the present moment*. She does not, cannot, rest peacefully in the Love that she here and now receives. She cannot enjoy anything at hand (Letter 8, 27ff.; Vision 13, 179ff.). Why? Because of that with which she fills and rules her consciousness. Again, think of devout medieval women before the Eucharist or the Crucifix.

(d) What Hadewijch positively does is simply, by conscious free choice, *to fill and dominate her consciousness by unsatisfied desire*, by her longing, felt need and passionate wanting! Hadewijch may have got courage to recognize her convulsing, unslaked need because she read it described vividly by Richard of Saint Victor (1957, 219–33; cf. introduction by Clare Kirchberger). But Richard accepts and praises that the fourth and highest degree of love (unsatisfied love) stays in one's earthly life unsatisfied. This lack of satisfaction is appropriate, he insists. Agonizing as the frustration of this violent love is, it is like Christ's own earthly love and eagerly does good to others as it suffers its agony of exiled frustration. This burning, frustrated desire yields its own satisfaction to the lover, as, for example, Bernard of Clairvaux and later the author of the *Cloud of Unknowing* attest.

In many places, Hadewijch takes the same stand on this unslaked longing by human lovers as Richard of Saint Victor, Bernard and other theologians of the twelfth century do. Here, as was seen above, Hadewijch describes and endorses and at times exhorts herself and others with unsatisfied love to resign themselves humbly or reasonably to frustration and go, like Christ, to serve others lovingly in suffering. But in the passages on human lovers' conquest over God, especially when she speaks of unfaith, Hadewijch takes a contrary stand. She wants to be with Love *now*! She cries out for it. She calls to Love, "Come now!"

(e) Even in Hadewijch's passages on human conquest over, and unfaith in, Love, her interpretation coincides with Richard's and Bernard's that their volcanic human desire for love is God's own way of loving, shared lovingly with them, moving mightily in them. Yet, as for Richard and Bernard, for Hadewijch this is still *her* desire. She does not experience it as a "being moved" or "being swept by" Love. At this moment the lover does not cry out to Love: "I am moved by desire for you!" The lover cries: "I want you!"

True, Hadewijch is responding to a demand by Love Itself that she call constantly for fruition and be in desire as great as Love. But it is *her* response to engulf Love with her desire so that "all God's artifice can not keep her from Love" (Vision 13, 179ff.). The desire is experienced by both Hadewijch and Richard of Saint Victor as at the same time their own *active self-involvement*. Hadewijch, as we saw above, understands her conquest over Divine Love *to come through her "will" or "free-choice"* as well as by God's choice.

If Hadewijch were simply seeking God's kingdom and her salvation, as those Guerric speaks to and of, then it would still be active, freely chosen involvement, acknowledged and even required by traditional, nonpredestinational, medieval theology such as that of Thomas Aquinas. Hadewijch's free choice would be possible only because of God's prior free, eternal commitment to save her and all others if they freely chose to believe in him and obey his commandments.

Hadewijch also follows traditional Christian theology in her understanding that the greatest active self-involvement for humans is a fusion of passion and free will. In the tradition, it is, at best, spiritual passion that the individual by free will makes fully his or her own (see Dreyer 1989). As we have traced out, and will more extensively later, the passion of Hadewijch that she freely joins has many marks of the bodily rather than the spiritual. My present point is that what conquers God is the freely willed passion of Hadewijch. The value of this love lies in its being both passionate and willed.

This conquering human love of Hadewijch is an ideal unacceptable (if conceivable) in much modern philosophy and theology, as that of Kant or Sartre, precisely because the intrinsic worth of her love lies inextricably in its being both "passion" and "free choice." Human exercise of the categorical imperative is free from passion. It is of little import to Sartre whether or not his existentialist heroine in *Is Existen-*

*tialism a Humanism?* chooses to follow out her passion. All that matters
is that she chooses and acts according to her choice.

This supreme desire, extolled by Richard and Hadewijch and
freely engaged in, is an unchosen primal reaching and opening by the
individual human being. As a result, this desire is an opening and
reaching that are of themselves for this individual at this moment a
convulsive, self-determining passion. The two aspects, the convulsive-
ness and the self-determination, i.e., the passion and the free willing,
of the opening and reaching are contrary, not contradictory, aspects of
the desire. They are aspects that complete and support each other into
fuller, more whole loving desire. In this, however, Hadewijch resembles other medieval thinkers, such as Thomas Aquinas and Bernard of
Clairvaux.

(f) What neither Thomas, nor Bernard nor any other traditional
Christian theologian could accept was that Hadewijch claimed by *her
willed, passionate desire to affect God, to cause God to respond.* Divine Love,
she said, could not resist her desire. It had to counter longing with
longing. It could not stay away. Hadewijch celebrating her triumph in
unfaith over God is unprecedented in Christian theological tradition
in that she rejoices in herself as a strong agent. As we observed earlier,
she could conquer Love because she had been conquered by Love and
in order to be further conquered by Love. Thus in her conquering she
interacted powerfully with God in a certain desired and willed mutuality and even equality.

"Proud!" would traditional Christians say, condemning her.
Hadewijch agreed with the epithet. She is proud of her love and her
strength in it. "God granted me to know the perfect pride of love; to
know how we shall love the Humanity in order to come to the Divinity,
and rightly know it in one single Nature. This is the noblest life that can
be lived in the kingdom of God" (Vision 12, 188f.). If Divine Love could
not resist Hadewijch's prideful power of love, mainstream Christian
theologians could and did. Christian theologians, with occasional
exceptions such as process theologians of our time, refused anything
like Hadewijch's account of basic interactivity and interdependence
with God. They did so because they refused to qualify God's self-sufficiency. As a result, a never-easing burden of Christian theology over
two millennia has been to struggle to reconcile the thesis of absolute
divine self-sufficiency with various Biblical passages, the free creation
of the world, the blessed Trinity, human free will, and so on. Consider,
for example, the awesome feat of conceptualizing three divine "persons" who each relate to each other, know each other, and love each
other without acting on, depending from, affecting, or being affected
by each other, for they have only one mind and one will between them.

Hadewijch may well be thinking along a current of Neoplatonic Christian thought that never entered mainstream orthodox theology. Bernard and other twelfth-century theologians seem to dally at times in this stream when they comment on the Song of Songs or the Psalms. Their metaphors seem to imply that the Divine Lover is interiorly affected by the bride, though in their metaphysical explanation of it they deny it. They imitate Pseudo-Dionysius, who wrote around the turn of the fifth century and influenced widely subsequent Christian thinkers in both the East and West. Pseudo-Dionysius says that God was rightly called "Yearning" [*eros*] and:

> we must dare to affirm (for 'tis the truth) that the Creator of the Universe Himself, in His Beautiful and Good Yearning toward the Universe, is through the excessive yearning of His Goodness, transported outside of Himself in His providential activities toward all things that have being, and is touched by the sweet smell of Goodness, Love and Yearning, and so is drawn from His transcendent throne above all things, to dwell within the heart of all things, through a super-essential and ecstatic power whereby He yet stays within Himself. Hence Doctors call Him "jealous" because He is vehement in His Good Yearning toward the world, and because He stirs men up to a zealous search of yearning desire for Him. (1940, 106)

Hadewijch steps beyond this tradition when she boldly and unequivocally spells out the interaction between Divine Love and herself, in which she by her free will and personal passion "touches" Love and "draws" Love to her.

In the mostly implicit constructive dimension of the present book I attempt fundamental ethics, not theology. Significant for my inquiry is that, in the experience of Hadewijch, divine life, the model for human life, is not pure self-containment: neither pure self-thinking nor pure self-loving nor indistinguishable oneness. Divine life includes a mutual affecting and interactivity with other, human individuals. This is a different model from the traditional one.

We saw earlier how the final, supreme union of God and a human lover is itself a mutual affecting. The end, therefore, and aim of all human life, what every human being truly wants and what alone can satisfy it, is mutual affecting with another Person. We see now how this is true, too, of the penultimate act by which the human lover gains that union. Just before the best, noblest, most satisfying moment in human life the human being is already, as an equal, acting on and affecting another Person as that Person acts on and affects him or her. The human being has thus a relative autonomy, a relative self-sufficiency, too. The individual can be proud of what he or she has achieved.

(g) The mutuality is deeply intersubjective. How does Hadewijch affect Love so powerfully that Love must come to Hadewijch and join with her? By Hadewijch's desire, we were told. Hadewijch's freely willed, overwhelming, unrestrained, desperate, unbelieving, all-absorbing desire to be with Love compels Love to, in return, desire even more to be with Hadewijch. What draws Love irresistibly is what irresistibly draws Hadewijch: the longing of the other to be with the one. This is the sweetness that each cannot resist and choose freely not to resist.

What if this be a picture of the most basic aim of human life? Well, then, my nature and being and life, inasmuch as they be human and good, are basically neither drawn only to contemplating truth or beauty or to thinking or to merging in perfect oneness nor, ultimately, to being a successful explorer, scientist, scholar, artist, capitalist, and so on. Nor is it enough to say that my nature and being and life are drawn to loving love and identifying with it. I am drawn most basically to being wanted and eventually enjoyed by others while I want and eventually enjoy them. This is the "knowing" and "loving" to which I am most fundamentally drawn. It is— *pace* Thomas Aquinas and Bernard of Clairvaux, Immanuel Kant, G. W. F. Hegel, and Jean-Paul Sartre—the highest and greatest, as well as the most satisfying experience that a human being can live.

If Thomas Aquinas had been aware of his contemporary, Hadewijch's theological understanding of the summit of human life and experience, he might have observed, "She has interpreted the spiritual in terms of the bodily." In his exposition of passion Thomas affirms and applies as an operating principle that the more spiritual a passion is, the more independent it is of other persons. The more dependent an individual's passion is on another reality, the more bodily. Since God is pure spirit, we can speak of God feeling "passions," only if we understand that he is completely independent of any other reality in feeling them. In their passions as in everything else, angels are completely independent of each other, whereas humans, being bodily, depend on other humans in passions as in everything else (ST I–II, 22ff.).

Hadewijch would not have called "bodily" the mutual conquest of Love and herself of each other, but neither does she call it "spiritual." The mutual conquering that she describes between her and Love mirrors patently, consciously on her part, the reciprocal interrelating of embodied desires of two human, romantic, sexual lovers, illustrated abundantly by courtly love stories. It mirrors also, as Hadewijch intimates in places, the mutual, embodied loves of mother and child.

(h) There is yet another characteristic of Hadewijch's conquest over Love. When Hadewijch has expanded her desire as much as she can and strains in frustrated agony as Love still stays away, she does not always, as Richard of Saint Victor does with his parallel desire, return to earthly life to carry out this love in suffering service to others. At certain times Hadewijch speaks out her love to Love. She says, "I love you so much. I don't believe that you love me. I seem to be the only one loving. I fear you do not love me at all. I fear you do not help me" (e.g., Vision 13, 179ff.; Letter 8, 27ff.).

As Hadewijch tells it, she and other elite lovers in their time *turn the battle by thus voicing their unbelieving desire for Love.* In Western thought down through Hadewijch's time, the "word" (*logos* or *verbum*) has always been crucial in humans' attaining the final goal. But it has been God's own word, either as uttered by him or repeated by the human mind and voice. I know of no other Western thinker up through Hadewijch's time for whom the final, decisive step to supreme loving union with God is *a word, proper to the human God-seeker,* a word that God or Love or Jesus Christ does not say and could not say. (A God-Man could not say to God: "I do not believe you!") When the human lover speaks this kind of word to Love, she engages in real dialogue. Her *interpersonal speech* is what opens the totality of God to her She only speaks and offers no promise or account of deeds as a knight might to his lady love.

Hadewijch's words of unfaith express freely chosen, overwhelming, all-absorbing, unrestrained desire for Love. She makes clear that she is practically nothing but willed desire. This is what moves Love to come to her. It is this desire whose sweetness Love cannot resist or stay away from. But note that it is not Hadewijch's desire alone that moves Love to come to mutual embrace with Hadewijch. Love knows of Hadewijch's swelling desire and stays away until Hadewijch speaks about it to Love. It is Hadewijch's *voiced* desire that Love cannot stay away from. It is voiced desire that opens God for Hadewijch. Is this perhaps a first in medieval theology? Is it a first in medieval piety? Is there any mystic or holy person before Hadewijch who by speaking out her desire makes God come to her simply to be with her?

In any case, Hadewijch has, again, described her conquest over Divine Love with a characteristic that is typically bodily. Her voice is not the desire and unfaith which it expresses. It is the expression of them through sensorily perceived sounds. Even if Hadewijch intends "voice" only analogously, she means something "like" what is uniquely bodily. As Thomas Aquinas points out, spiritual beings, such as angels do not have anything like spoken words. Angels cannot communicate to each other directly but do so only by God's mediating causality. There is no place in traditional Christian theology for something spiritual that would be itself neither an act of thought nor will but the expression of such an act. When a spiritual creature "speaks" to God, it means only that the Creator, conserving them in existence and concurring in all their act, knows all their thoughts and willings. When a spiritual creature wants to speak to another spiritual being, God inserts the pertinent cognitive *species* in the intellect of the other (ST I, 54-58, 107).

This leads us to the second part of the book. In Part I, we noted that the intersubjective mutuality which Hadewijch, contrary to Western traditional thought, makes the supremely worthy, all-fulfilling aim and ideal of human life is of a distinctively "bodily" kind. It is bodily in that it has traits that traditional theologians considered peculiar to the bodily. It is bodily, too, in that it has traits that evoke for any reader, medieval or modern, analogy with traits that are common to what the reader considers bodily.

In Part II, we scrutinize with some analysis and synthesis *what* these characteristics may be. We listen not only to Hadewijch but also to other devout women of her time and place. What, more exactly, are some bodily aspects of the intersubjective mutuality which Hadewijch and other devout women of her time and place, cherished as supremely precious and satisfying? In Part II, we focus our gaze on the *knowing* dimension of this mutuality, where in Part I we centered on the *loving*.

## PART II

Medieval Women and Bodily Knowing

# 8

In Part II we start over. While recording in Part I traits of the mutuality that Hadewijch prized for constituting the best of human life, we noted that the mutuality was consistently embodied. We noted that also many other women of Hadewijch's time and place prized this embodied mutuality. Now in Part II going back and beginning afresh, we attend to this embodiment that, the women say, characterizes supremely worthwhile moments of human life. These are generally moments of mutuality, too, but our effort now is simply to discern in what sense they are "bodily." Can we trace recurrent lineaments of the bodiliness of such experiences of these women? To enable more intensive analysis, in Part II we center our gaze on the *knowing* dimension of this bodiliness whereas in Part I we centered on the *loving* dimension of the mutuality.

Again, a resemblance, at least a surface one, with late twentieth-century women writers is patent. In the last decade or two, a number of women writers make the point that most, if not all, good human knowing is embodied. The claim is, as I am about to document, not unprecedented in Western thought, but contemporary women make the claim plausible and urgent in their chilling critique of male-formed Western culture. The claim is made, for example, by Naomi Goldenberg (1989), Susan Griffin (1978, 1982) Carter Heyward (1989), Mary Hunt (1991), Audre Lorde (1984), Judith Plaskow (1991), Adrienne Rich, (1976), Rosemary Ruether (1983) and Haunani-Kay Trask (1986).

The claim is twofold. Embodied knowing can be valuable in two respects. First, most knowing that is *worth having for its own sake* is embodied. Instances of bodily knowing, such as the mutual awareness of friends together, are among the things in life most worthwhile in themselves. Lives of Western civilized persons are impoverished because dominant trends in the culture refuse to endorse and encourage such bodily knowing.

Secondly, *to know in given situations when and how to act* requires embodied knowing. One must first know in embodied experience the human realities pertinent to the situation and their relative worth and importance. The women score the neglect of bodily knowing in various areas of human decision and action. The areas include war and peace, social justice, reproduction and sexuality, environment, education, and interpersonal relations in general. To decide and act well, one has to know how valuable in the situation are the realities at stake. In modern Western culture, these realities are wrongly or inadequately assessed because, among other things, male minds continuing to shape the culture use only a disembodied, "rational" mode of knowing.

I have learned from this critique. I want to learn more from it. The women, however, whom I have read so far do not provide all the intellectual articulation I look for. Even less do the few male authors I know who argue, in conscious opposition to their Western tradition, for the intrinsic value and importance of the bodily in human life. From these men, too, such as Don Browning (1983), Andre Guindon (1986), Leon Kass (1985), Sam Keen (1970), Daniel Maguire (1978), and James Nelson (1978, 1983, 1988), I have learned but look for more. Moreover, In reading some of these men and women I cannot escape the impression that in articulating values of embodiment as in articulating some of mutuality, the twentieth-century authors have not freed themselves completely from the confines of male establishment thought. As I said in the Introduction, I do not argue this explicitly, but I suggest throughout the book that in some respects medieval women give a fuller account than late twentieth-century men or women do of the riches of embodied mutuality possible in human life.

On the other hand, if I read more of the corpus of the contemporary authors cited above, I might find the intellectual articulation I want. As it is, however, I pause, ponder, and respond. In what I have read, these thinkers, male and female, do not examine systematically what is "the bodily" to which they refer. The word "bodily" has, after all, many meanings. What do they mean by it? Wherein lies its peculiar value? My own efforts in this regard are not that pellucid on rereading.[1]

In Part II, therefore, I take as given that we humans have at times a kind of bodily knowing that is itself uniquely worthwhile, an end in itself. I take as given that sometimes by this kind of embodied knowing we know in a unique way what is truly good. My question is: What is such knowing? What is this precious kind of bodily knowing? How do human beings know values through their bodies? In what sense bodily?

My question may reflect a male mind's need to do the impossible, to reduce the nonrational to the rational. But it seems pertinent—I nearly said "reasonable"—to ask what distinguishes worthwhile bodily knowing from nonbodily knowing that may also have its own worth and use. The question echoes Socrates' relentless "What is…?" (*ti esti*). But Socrates did not ask it of anything bodily. Nor did he prize bodily knowing.

I look then for an idea of intrinsically valid and valuable bodily knowing. Nothing so ambitious as an idea that encapsules the bodily knowing itself or even represents it. I look for an idea that in its articulation points to, or helps identify, in experience this precious kind of knowing. In other words: What might be a starting list of recognizable traits of this bodily knowing? The list could help to locate the experience so one could move in closer to try a more precise and extensive phenomenology. On the basis of the more advanced phenomenology, then, one might also determine more exactly what this bodily knowing knows and where and how it can be had and further developed.

In Part II, therefore, I try only to start such a list. I propose it as a tentative identification of some bodily knowing that is worthwhile in our lives. In the following pages I argue boldly for my list, but in truth I aim simply at advancing grounds for a hypothesis that would invite discussion, debate, and—above all—further testing in experience.

I derive the list from women's experience in the twelfth through fourteenth centuries in northwestern continental Europe. I am indebted greatly to social historian Caroline Walker Bynum, art historian Joanna Ziegler, and psychological historian Marilyn Mallory. I derive the list also from Hadewijch's writings, whose experience we concentrated on in Part I.

My concern—to say it once more—is not medieval women's religious knowing as such but what I find in their religious knowing that may be true of all human knowing, religious or secular. At each trait on the list I will offer inchoative analysis and phenomenology to suggest how this medieval experience is, for all the differences, analogous to experiences we have today and how it seems to qualify today, too, as worthy bodily knowing. "Bodily" may not turn out to be the most felicitous term for what I am trying to identify. If so, consider it an agreed upon label or codeword which I use in constructing my hypothesis.

In the Middle Ages, as in the late twentieth century, women urged on the dominantly male Western culture a bodily way of knowing. But in the Middle Ages, as in our time, the picture is far from being all or nothing. Christian medieval piety as a whole was more bodily than ever before in Christianity. It became more and more bodily as the Middle Ages progressed. Still, as Bynum and Ziegler point out, this "physicality" of devotion was found more in women than in men.

Bynum stresses the consonance between this trend of popular piety and the contemporary development of medieval theology. The theologians, mostly male, came in the same period to give more credit to the bodily, as they did to humanness in general. Bynum affirms a parallel progress of the male theologians and the devout women and suggests a mutual influence (Bynum 1984, 199–201; 1985a, 12–13).The parallel, however, between the women's devotion and the theologians' ideas falls short. The women's religious experience reported by Ziegler and Bynum outran the theology. Bynum notes some of this divergence but not as much as I do in the present book. In the extent of its physicality, the experience contradicted the theology. The women experienced as bodily knowing and prized as real knowing what theologians generally ignored or denied could be real knowing. This is historically typical and significant, for in Christianity, more than in Judaism or Islam, the thinker, i.e., the philosopher-theologian, is a conspicuously integral part of the public community. The theologian influences massively and is massively influenced by the culture and experience of the whole community. In the Middle Ages this mutual influence flowed back and forth in the training of and teaching and preaching by clergy, monks, and some exceptional women like Hadewijch.

In the Middle Ages, as in succeeding centuries, Church theologians generally treated religious experiences of this physical, popular kind as delusion or, at best, pious imagining. It was not a knowing of anything important. The theologians did not criticize the official toleration and popular enthusiasm for such experience. They did not contest that the experience might be genuinely pious and, under supervision, might have good spiritual effect on the faithful. But by ignoring or by cautioning they treated this physical devotion as essentially subjective. A modern atheist has to treat the women's religious experience as subjective because the experience was religious. Medieval theologians treated the women's experience as subjective because the experience was too bodily. Why did they have to?

The reason for the theologians' recalcitrance is not immediately evident. By the thirteenth century, theologians, aided by Aristotle, were affirming more strongly than ever the goodness of the human body and its indispensable, positive place in human life. A human being did not live in his body; he was his body as he was his soul. He had bodily knowing, i.e., sense knowledge, distinct from rational knowing. His rational knowing depended from, corresponded to, and was limited by his bodily knowing. Why then did the theologians deny, or refuse to affirm, that the kind of bodily experience that the women reported could be true knowing? What in the experience led them to reject it? What characteristics of the experience which the women claimed to be genuine knowing compelled the theologians to treat it as pious fantasy?

This is the enterprise of Part II. What traits of the medieval women's experience could the theologians not accept as characteristic of true knowing? What unacceptable traits emerge clearly in that the epistemology of the theologians rejected them in principle while the women described them in flesh? My main work will be to draw up a list of these traits. The listing, however partial and superficial, can form, as I have already suggested, a usable focus for contemporary probing of the nature and value of bodily knowing, whether in a religious or secular context.

Neither Bynum and Ziegler nor I claim an absolute division between women and men in the respects in which we contrast them. Not all medieval women were affected significantly, or at all, by the embodying trend of medieval spirituality. Bynum describes how the physicality of devotion which she finds widespread in medieval women is verified for some medieval men. She adds,

> Moreover, careful reading of the texts produced by male mystics reveals that those whose piety comes closest to that of their female contemporaries were invariably involved with advising communities of nuns, were usually deeply influenced by their own mothers, and were inclined to describe themselves as women. Bernard of Clairvaux, Francis of Assisi and Henry Suso, for example, all preferred female images for themselves. (Bynum 1985a, 6; cf. Bynum 1987, chap. 3 and 10)

Bernard McGinn has drawn my attention to evidence in some male medieval thinkers, especially mystics, for the kind of "physical knowing" I characterize in the present essay. McGinn offers as illustration a description by Rupert of Deutz (c. 1075–1129) of a dream vision he had of the living Christ. "I took hold of him whom my soul loved. I held him. I embraced him, I kissed him for a long time....In the midst of the kiss he opened his mouth so that I could kiss more deeply."[2]

I noted above a few male ethicists of the late twentieth century who join women in appreciating bodily knowing. Certain male medieval theologians openly esteemed and recognized some bodily knowing and influenced women writers in this respect. The devotion of Bernard of Clairvaux to scenes of the life of Jesus is a striking example. Some modern women seem to me to think principally along the exclusively rational tracks of the dominant culture. Nevertheless, on the whole, a similar tension seems to hold roughly in the late twentieth century in this country and in the thirteenth through fifteenth centuries in the Low Countries and Germany. In both settings, many women prize bodily knowing to an extent which contemporary male thinkers, even a Bernard of Clairvaux, generally do not. I will describe this tension in its medieval setting and suggest its analogy to what occurs today.

In medieval and modern times, looking at this kind(s) of knowing that the women experienced and prized suggests that the men were not only wrong in depreciating such knowing but that they misunderstood its nature. If so, it would follow that they misunderstood also the nature of the kind of knowledge, "rational" or "intellectual" knowledge, which they did endorse and champion. In some sense, as they acknowledged, rational knowing depends on prior bodily knowing. If they misunderstood the latter kind of knowing, then they are likely to have misunderstood the dependence that rational knowing had on it. They are likely then to have misunderstood at least some of the nature of rational knowing. These suggestions are not taken up in the present inquiry, but have influence on the questions asked.

My reflections were set in motion by an exhibition of medieval sculpture at the College of the Holy Cross, entitled "The Word Becomes Flesh," curated by Joanna Ziegler. My thought was stimulated further by the connected symposium, organized by Ziegler, entitled "The Word Becomes Flesh: Radical Physicality in Religious Sculpture of the Later Middle Ages." The pieces of the exhibition were examples of the extremely physical kind of sculpture commonplace in the Low Countries in the thirteenth, fourteenth and fifteenth centuries. In the introduction to the exhibition's catalogue and in her paper at the symposium, Ziegler argued that the art responded to a demand of the women's religious movement of the time. The "radical physicality" of the art reflected the physicality of the women's piety (1985a, 9–37; 1985b; 1990).

Where earlier art had generally portrayed the joy, glory, and triumph after death of Christ and his followers, this Low Countries art, like much Rhenish art of the time, tended to portray moments of their thisworldly life. Far more than artists of Christendom had ever done before, the Low Countries artists expressed "physical and psychological states of being" recognizable by their viewers then as today (Ziegler 1985a, 13). The artists did so by force of physical detail. The pieces of the Holy Cross exhibition assault the senses of the viewer with the agony of Jesus in his passion and death or with the responding anguish of Mary, John, and other followers: a twisted grimace beneath the thorns, sagging shoulders beneath the cross, open mouth of horror, and spread arms and legs embracing the foot of the cross. (1985a, 39–61). The reader may recall a passion scene of a Grunewald or Riemenschneider.

Walking from one wooden figure to another in the Holy Cross exhibition, one was—stunned? shocked? drawn to? repelled?—by the agony portrayed. It was hard to avoid feeling some of the feelings of the figure. It was hard to avoid some "physical and psychological" feeling of one's own, whether empathetic or antipathetic. This is another novelty of the new art. Not only did it depict familiar bodily feelings, but it tended to involve the spectator in physical, emotional response to the feelings of the figures.

All art—Ziegler cites Huizinga—is applied art. The market for an art shapes the art. Much of the market shaping this art, Ziegler argued, was devout women of that time and place.

> With the thirteenth century the women's religious movement constituted one of the largest and most serious components of the new piety, particularly in Germany and the Netherlands....In fact, it would hardly be an exaggeration to say that women were the dominant feature along the Rhine and in the Low Countries. (1985a, 19)

The new devotion of the women, in particular their devotion to the Eucharist, created the demand for the art. Ziegler analyzed the art, particularly the original art form of the Pietà, to suggest how the physicality of the art reflected the eucharistic piety of the women (1992).Ziegler supported her conclusion by her research into not only art but also written records of the life and eucharistic piety of women of northwestern Europe from the thirteenth to fifteenth centuries. She concentrated on the communities of lay women called Beguines, which were then found throughout the Low Countries, and invoked convergent research by Caroline Bynum (1984, 1985a, 1985b, 1987).

Bynum recorded evidence of various sorts to show how the piety of women of the late Middle Ages was more physical than any Christian piety before it. The women's devotion epitomized this. Unprecedented in Europe was the eagerness of the faithful, especially women, to receive the Eucharist, to eat of the bread, drink of the wine, and thus receive Christ within them. Similar was their eagerness to venerate the Eucharist when consecrated bread and wine were placed in a tabernacle of the Church and they could pray before it. The Church during the late Middle Ages made possible much more frequent reception and veneration of the Eucharist than hitherto. The ardent desire of women was a principal cause of this change.

Devotion to the Eucharist is by its nature physical, for its reception involves eating and drinking with the belief that Christ is then in one's body. Other veneration of the Eucharist involves cherishing physical proximity to Christ in the tabernacle. But the rapturous experiences of the women that Bynum retailed in relation to the Eucharist were, as Ziegler's art, of an even more physical piety, even more unprecedented in Christendom in its physicality. The raptures and ecstasies were numerous throughout northwestern Europe at the time and of enormous popular interest. The intense interest of women who may not themselves have directly had such experience indicates a similarly physical piety.

In the eucharistic raptures of some women, the wounds of Christ appeared spontaneously on their hands, feet, sides, and faces. The reception of the Eucharist led so naturally to stigmata that contemporaries hardly worried about how to account for their appearance. Women who did not experience such mystical or paramystical phenomena felt similarly, though in a more moderate way. "No religious woman failed to see and feel Jesus as wounded, bleeding and dying" (Bynum 1984, 189). Women desired in prayer to take the dead Jesus down from the cross onto their laps. Ziegler concluded to a similar devout desire in women when explaining the origin and popularity of the sculpted Pieta (1985a, 12–25; 1992).

The physical union which the women sought and had with Christ was not only that of sharing his agony and passion. They saw often in the host and chalice Christ the baby or Christ the bridegroom and they joined him.

> Agnes of Montepulciano and Margaret of Faenza became so intoxicated with the pleasure of holding the baby that they refused to give him up. Ida of Louvain bathed and played with him. Gertrude of Helfta, Lidwina of Schiedam, and Gertrude of Delft nursed him at their breasts. Angela of Foligno and Adelheid Langmann married him in the Eucharist. (Bynum 1985a, 9–10; cf. 1984, 188–89)

Margaret of Oingt writes: "And in the evening when I go to bed, I in spirit put him in my bed and I kiss his tender hands and his feet so cruelly pierced for my sins. And I lean over that glorious side pierced for me" (Bynum 1984, 191; cf. translation by Renate Blumenfeld-Kosinski 1990, 63; Ziegler 1985a, 24).

Bynum notes that the physical union experienced by some of these women "sometimes culminated in what appears to be orgasm—as in Hadewijch's beautiful vision (1984, 191). Hadewijch, as we heard earlier in this book, recounts:

On a certain Pentecost Sunday I had a vision at dawn. Matins were being sung in the church, and I was present. My heart and my veins and all my limbs trembled and quivered with eager desire and, as often occurred with me, such madness and fear beset my mind that it seemed to me I did not content my Beloved, and that my Beloved did not fulfil my desire, so that dying I must go mad, and going mad I must die. On that day my mind was beset so fearfully and so painfully by desirous love that all my desperate limbs threatened to break, and all my separate veins were in travail....I desired to have full fruition of my Beloved and to understand and taste him to the full....

After that [her reception of the Eucharist], he came himself to me, took me entirely in his arms, and pressed me to him, and all my members felt his in full felicity, in accordance with the desire of my heart and my humanity. So I was outwardly satisfied and fully transported. Also then, for a short while, I had the strength to bear this; but soon...I saw him completely come to nought and so fade and all at once dissolve that I could no longer distinguish him within me. Then it was to me as if we were one without difference....After that I remained in a passing away in my beloved, so that I wholly melted in him and nothing any longer remained to me of myself. (Vision 7)

Hadewijch, as we saw, describes similar passion and union in Visions 9 to 14. She expresses it generally in a letter:

Where the abyss of his wisdom is, [God] will teach you what he is, and with what wondrous sweetness the loved one and the Beloved dwell one in the other, and how they penetrate each other in a way that neither of the two distinguishes himself from the other. But they abide in one another in fruition, mouth in mouth, heart in heart, body in body, and soul in soul, while one sweet *divine nature* flows through both and they are both one thing through each other, but at the same time remain two different selves—yes, and remain so forever. (Letter 9)

As Ziegler and Bynum stress, theology kept pace to some extent with the physicality of the new eucharistic devotion and new religious art. Each of the three—theology, piety, and art—brought out the humanity of Christ as never before in Christendom. Each in its own way gave Christ's body more importance than ever before. Each, too, made the Christian's body more central to Christian life. For example, while frequent reception of the Eucharist became more and more popular and art became more and more realistic in bodily detail, Thomas Aquinas, following the lead of Albert the Great, introduced Aristotle's hylemorphic anthropology into hitherto Platonic theology. In a variety of other respects, and in other media and milieux of Europe in the late Middle Ages, the physicality of Christ and Christian advanced upstage.

But in medieval theology, as I stress, the advance was limited. In fact, this traditional theology, marching past the women down the modern centuries to the present day, has still not assimilated their physical experience. I know no contemporary Christian mystical theology offered as a continuation or extension of Christian tradition that integrates into its system or endorses in principle the physical experiencing of Christ such as medieval women tell of and as exemplified above. Many Christian thinkers deny that such experience can be anything but delusion or pious imagining.

For traditional Christian theology, all experience of God must be noncorporeal, whether the experience be by intellect or by loving will and its affects or even beyond will and intellect. Mystics and contemplatives cannot taste the sweetness of Jesus through their body. Their mind must move "higher" or "deeper" to some nonphysical level in order to find and savor this sweetness with the mind's higher faculties. The necessity of rising above all bodily experience, sense images, and bodily passion in order to come to any experience of God is repeated monotonously in mystical theologies, including those of Origen, Augustine, Cassian, Gregory the Great, Pseudo-Dionysius, Maximus Confessor, Bernard of Clairvaux, Hugh of Saint Victor, Thomas Aquinas (Hodgson 1944, liii, lix–lxi, lxv–lxxi, lxxv; Johnston 1973, 24–31; Butler 1927, xxxiv, lxxvi–lxxviii, 183–88). Thomas Merton stresses similarly the necessity of leaving behind all visual images (1977, 110–11).

This necessity is affirmed by both main currents of mystical thought in Christian history. The apophatic theology of Plotinus holds sway in the East from the earliest centuries and in the West from the medieval period on after Pseudo-Dionysius was translated into Latin. A more sober, less negative current of mystical theology dominates the West during the earlier centuries through the work of Cassian, Augustine, Gregory, and Bernard. In both theologies the mystic must leave behind all sense and bodily experience to attain an experience wholly different (Butler 1927, 179–92). Many women visionaries echo this verbally in their own theology while they assert also the contrary experience. For instance, Hadewijch, as noted in Part I, describes her supreme union with God as out of spirit and out of body.

True, the tradition recognized that some physical experience might precede, accompany, or follow the purely spiritual rapture of the mystic. Some medieval thinkers, such as Bernard of Clairvaux and others influenced by him, even extolled such pious imagining. However, not only is this imaginative devotion of limited value, to be, when possible, transcended into the purely spiritual, but also the physical, actual imagining of Mary and Child at Bethlehem serves only as context for spiritual devotion.

The great twentieth-century philosopher of mysticism, Evelyn Underhill, continues traditional thought when she explains what part the physical can play in an individual's experience of the divine and what part it cannot. The experience of God, the union itself with the divine, cannot be physical. The union must be nonbodily, spiritual.

"Rapture," defines Underhill, is a mystic's state of bodily trance with drastically reduced bodily perception. Underhill quotes Thomas Aquinas:

> The higher our mind is raised to the contemplation of spiritual things, the more it is abstracted from sensible things. But the final term to which contemplation can finally arrive is the divine substance. Therefore the mind that sees the divine substance must be totally divorced from the bodily senses, either by death or by some rapture. (Underhill 1930, 361; she quotes *Summa contra Gentiles*, III, xlvii, Rickaby's translation.)

The word "rapture" in English has lost much of the meaning of the Latin *raptus*, as used by medieval theologians. *Raptus* means literally "being carried off by force." In contexts other than mystical experience *raptus* can mean "robbery," "plundering," "abduction," or "rape." For Underhill, drawing on numerous accounts and interpretations of mystical experience, the impact of the union with God is to carry the soul away from the body. There is a withdrawal of life and energy from the body, for the union itself is purely of the soul.

The physical serves at most to "interpret," "translate," or "express" the spiritual experience. In rapture, the body of the mystic can be stunned by the intensity of what is going on in his or her spirit. Physical experience is then an overflow, a secondary accompaniment, of the spiritual. This is the most it ever is. Underhill echoes here the views of late medieval mystics, Walter Hilton and the author of *The Cloud of Unknowing* (Underhill 1930, 266–70, 359–79; Colledge 1961, 60–62; Hodgson 1944, lxxi). Thomas Aquinas explains similarly how the beatitude of the heavenly blessed, though it consists primarily in spiritual contemplation of God, has also a bodily component: "in perfect beatitude the whole human being is perfected, but in the lower part through overflow [*redundantiam*] from the higher" (ST I–II, 3, 3; my translation).

As J. Aumann's article "Contemplation" in the *New Catholic Encyclopedia* summarizes, "The various grades of contemplative prayer produce definite reactions that are sometimes manifested even in the body" (263). All contemplation is of the speculative intellect, though using other cognitive powers as auxiliaries (258).

Illtyd Trethowan declares that extraordinary forms of mysticism are "sometimes accompanied by extraordinary physical conditions (levitation, for example), visions, and locutions. The competent authorities agree that such manifestations have nothing to do with the substance of mystical prayer and that they normally disappear before its highest stage has been reached" (79; cf. 86, 88).

A main thrust of Charles Davis's *Body as Spirit: The Nature of Religious Feeling* is to oppose a common rejection of the body by religious consciousness, both past and present, and to affirm that religious feeling and all truly human feeling is essentially both spiritual and bodily. Yet Davis—consciously, I believe—follows tradition in seeing the bodily as no more than the "medium" or "expression" or "participation" or "presence" of spirit. Any important knowing of reality, any "cognitive apprehension" of real value, is not of the bodily itself, but of the spirit that is embodied (4–13, 32–58).

The medieval women whom Caroline Bynum listens to say often the contrary of what these traditional theologians, medieval or contemporary, say. The women describe a certain knowing union that they have had with Christ. They accord it, simply in itself, the definitive worth and importance that the tradition attributed to contemplation or mystical union. The women report their union with Christ as bodily, but it is no two-level experience. They testify to a single experience of, for example, the suffering Christ or the sweet Host or Baby Jesus or Christ as lover. The women's physical experience did not express anything. It was itself. They held Christ physically. That was the whole thing. That was the wonderful, longed for, enjoyed, whole thing.

The women knew, of course, that their union with Christ was not, for example, normal, natural eating or lovemaking. They would have nothing to do with natural sex. They tended to eat little, if any, normal food. They knew that as they embraced God in prayerful rapture, their actual hands and arms may have been at their sides or folded in prayer. It was like dreaming or daydreaming or fantasy but for them it was a real knowing of something really happening. They knew this real happening as truly as Peter and Paul knew Jesus appearing and speaking to them. But what was important and wonderful in what was happening to them was not, or was much more than, a sense knowledge of the kind that mediated to the spiritual intellect of Peter and Paul a message of practical import. It was the direct bodily knowing and loving by them of Jesus.

The women took their experience of Christ as supernatural. It was a miracle, a gift from God. But the miracle, for example, was that it was, analogously but truly, an eating of him, nursing him, making love with him, or holding him dead. They possessed and experienced Christ through bodily perceptions, bodily interactions and bodily feelings, including holding, giving suck, eating, experiencing orgasm, and playing. They felt this were true even if they were not swept up in mystical experience, even if they were simply absorbed in ordinary Christian prayer, before, say, a Pietà.

It is not the intimacy of the women's union with God that repels the theologians. The theologians affirm with awe the intimacy of the soul and God which God makes possible on the mystical heights and, more completely, in the immediate vision of God that all the faithful have after death. It is the thoroughgoing physicality of the women's experience that make it unintelligible and often repugnant to mystical theologians of the tradition, even modern ones. Thus, Wolfgang Riehle notes:

> In comparison with [Mechtild of Magdeburg's image], Margery Kempe comes off very badly, for in her treatment of this motif she shows the unpleasant side of medieval female mysticism. In her book Christ thanks his "dowtyr" for so often having bathed him in her soul, and when he appeared to her in a dream she took his toes in her hand and felt them. This is certainly an expression of the theme of man touching God, but the "detour" via the idea of the incarnation of God becomes almost an end in itself for Margery. For her it is nothing short of the highest goal to be able to feel the physicalness of God *realiter*: "& to hir felyng it weryn [i.e., Christ's toes] *as it had been very flesch & bon....*" This process of "materializing" sensual imagery, which we can observe here in Hilton's translation of the *Stimulus Amoris*, is again something which was taken to extremes in texts of female mysticism. The great mystical texts are satisfied with the implied suggestion of mystical communication, but in the texts of the women mystics the imagination knows no such restraints....These images...stand out...because of their complete lack of any spiritual reference. (1981, 11–19)

Riehle asserts the superiority of Julian of Norwich to the German women mystics in that "Even when she has a corporal vision of the Passion of Christ this is not really an end in itself, but serves only as a starting point for some theological abstract knowledge" (126; cf. 114, 116). Riehle makes an important point: the women mystics in the northwestern Continent report often their bodily mystical experiences as ends in themselves. The question is: Why are theologians, like Riehle, sure that this is a mark of inferiority of the experience or its account?

In the Bible and in Christian tradition, physical experience can be a revelation from God. Jacob saw a ladder with angels descending and ascending (Gen. 28:10–22). Peter saw all kinds of animals and birds, walking, crawling, and flying (Acts 10:9–35). Christian thinkers understood that God sent at times physical images to tell spiritual truth. The physical serves to symbolize the nonbodily. Christianity always recognized that what a person visited by God perceives with his or her eyes or ears may express spiritual truth. New was the kind of eucharistic experience on which Bynum and Ziegler focus. Here, when the women are visited by God, somehow they experience Christ with their senses.

Christians have always compared their experience of God with bodily experience. But they were speaking metaphorically of what they believed to be a spiritual experience. Augustine inspired many to exclaim with him: "You [O God] called and cried out and broke my deafness. You flashed and shone and put my blindness to flight. You gave fragrance and I drew breath and now I pant for you. I tasted and I hunger and I thirst. You touched me and I burned for your peace" (*Confessions* X:27, my translation). But none could be clearer than Augustine that this kind of language can be only a metaphor for the spiritual. "You, O God, are sweeter than all pleasure, but not for flesh and blood" (IX:1; cf. X:6–7).

Sensual metaphors with which theologians described their relationship with God grew more abundant and more sensual as the Middle Ages progressed. The erotic imagery of Bernard of Clairvaux and other Cistercian monks of this time, collated perceptively by Jean Leclercq, is a case in point. But theologians who used them made clear that the images were only metaphors. The physical images stood for spiritual, nonbodily experience of spiritual, nonbodily reality (Leclercq 1979, 103–4).

Hadewijch, Margery Kempe, Margaret of Oingt, and other women recount often physical experiences as metaphors for their relationship with God or as allegories or symbols of divine truth. They know then that they are only expressing in imagination what they believe by faith. At other times, however, they tell of what they experienced, not as metaphor, symbol, or expression, but as fact: Christ himself really, physically interacting with them. Is it possible to read the lines of Hadewijch or Margaret of Oingt quoted above and believe she was simply using metaphor to express what happened to her?

Were such physical experiences of Christ all illusions? I cannot judge. Some of the experiences are bizarre to modern ears. Some were even bizarre to medieval ears, but some of these women were recognized by their contemporaries and by later historians to be wise, mature women, acting responsibly and effectively in the medieval world. They were consulted by theologians and other men, and much of their advice seems sound today. But this does not prove they did not hallucinate in their physical raptures or confuse fantasy with fact.

Whether their raptures, "mystical" or not, were illusions or whether the women truly experienced the real Christ does not affect my point. My point is that they evoked a specific kind of knowing, secular or religious, in describing their visions, etc. My point, therefore, is the nature of such knowing, its possible value. My point is emphasized by the response of the theologians. Or more accurately, their lack of response. Why did the theologians not acknowledge that rapture of this kind *could* be a true experience of Christ? What kind of experience was it that made it for them unacceptable, inconceivable?

To ask this question sociohistorically or psychologically might be more fruitful, but I, sticking to my last, ask it philosophically and phenomenologically. Did any element of their theology block these men from seeing the possible validity of the women's experience? More precisely, did anything in the theologians' theory of knowledge keep them from recognizing the human mode of knowing which the women exercised? That is, what properties of the women's experience did the theologians' epistemology exclude in principle as possible characteristics of true knowing? The answer that I am about to propose is not peculiarly religious. Closer scrutiny shows that the theologians' refusal does not stem ultimately from their understanding of human knowing of the divine. Their refusal stems ultimately from their understanding of human knowing of any person, human or divine. Their refusal stems, therefore, from their understanding of what constitutes any important, worthwhile knowing.

# 9

What obstacles could there be to the theologians' assimilating into their system the women's experiences? For traditional theology, God can take on the properties of any physical reality he wants. He became man. In the Eucharistic host he has the physically experience-able properties of bread. The Christian eats Christ, as Christians have said from the beginning, echoing John 6, 1 Cor. 11:23–24, and confessions of faith which Berengarius was compelled to make by the Council of Rome in 1059 and 1079 (Denzinger and Schonmetzer 1963, 690, 700).

True, the theologians insist that a person receiving the Eucharist does not literally taste Christ nor experience him physically in any other way. The Christian in eating the Host knows by his or her senses only the properties of bread and tastes only bread. Only by faith does the person know that it is the Lord he or she eats. But why could the theologians not acknowledge the possibility that the eucharistic Christ had on occasion by special grace enabled the loving recipient to experience him physically in his physical reality present in the sacrament?

An obvious answer is that the theologians knew in some sense a human being could know Christ by a bodily experience similar to what the women claimed. Contemporaries of Christ in his life on earth knew him in such a way. Some contemporaries, like Mary his mother, held him as a baby or as a corpse taken down from the cross, but, for the theologians, such a knowing of Christ is unimportant. Of itself it is not a worthwhile kind of knowing, unless the knower rise to spiritual knowing of Christ. For the theologians, all significant, intrinsically valuable knowing of Christ must be immaterial, noncorporeal. Why?

One reason is that for the traditional theologians, Christ is God, and as we saw, all experience of God must be immaterial, nonbodily. But why could the theologians not recognize that there might be also another way? Why could not Hadewijch in a single, fused experience of "body" and "soul" know "divine nature," as she claims in the passage cited above? Why could they not conceive at least the possibility

91

that chosen individuals, by God's extraordinary grace, might experience God in their very sensuality and physical passion?

The immediate answer is, again, obvious. God is pure spirit. He
is immaterial intellect and will. But why must God be pure spirit? The
divine nature is certainly beyond any spirit, intellect, or will that can
be known by ordinary human knowing. Might God not be loving
knowing that transcends our categories of body and spirit? Might God
then be knowable, however imperfectly, by either our bodily or our
spiritual knowing?

The reason for the theologians' refusal of bodily knowing of God
is broader and more fundamental than it first appears. One might
grant them, for the sake of argument, that humans could not experience in their body the divine nature. True, the women believed they
experienced more than a wonderful child or a wonderful man—more
than a saint, say. Their experience enraptured them because they experienced Christ as God as well as man. But one might simply fault them
benignly for theological inaccuracy. One might say they expressed
themselves poorly.

After all, few of the women claimed, as Hadewijch did, to experience God's divine nature. If they had put their conviction carefully in
traditional theological language, most of the women might have said
that they experienced Christ, the divine person, in his human nature.
They experienced the Incarnate God as the human being he also was.
Why did the theologians not work into their theology this claimed
bodily experience of the divine Christ in his humanity?

Here we see the broader, more basic refusal by the theologians.
Though he wrote in the thirteenth century, Thomas Aquinas represents many later theologians because Thomism spread widely in Western Christian theology before the Reformation and because, more than
the alternative Platonic, Neoplatonic, or Nominalist syntheses of the
late Middle Ages, Thomas's systematic accords and elaborates for the
body a central, estimable place in human life and knowing. It can be
argued that Thomas's Aristotelian epistemology is a high point in the
appreciation of bodily knowing by traditional Christian thought (ST I,
80, 2; I, 84–86; Miles 1981, 120–34).

The theologians insist that not only divine nature is immaterial.
God, angels, and humankind are all essentially immaterial, inasmuch
as they are all a "person," i.e., an "individual subsistent reality of rational nature" (*substantia individua rationalis naturae*). This is Boethius's
classic definition of person and Aquinas accepts it without question
(ST I, 29). Rational nature is essentially incorporeal. It cannot be bodily.

This is manifestly true in the Platonism which dominated Christian theology before the thirteenth century and in which the human

being was simply an immaterial soul. But even with the introduction of the Aristotelian understanding of the human being as a composite of body and soul, all that makes a person human as distinct from a brute animal is the soul in its purely immaterial reality. It is, therefore, for the same reason that one cannot know by one's body either God, angel, or human in their proper reality: they are rational (ST I, 75, 4–6; 76, 4; I–II, 3, 3; see also I, 80, 2; I, 12, 2). By this epistemology, bodily experience as such cannot attain the human being as such, the person as such. One must first perceive with the senses the individual before one, be he God, angel, or human. But according to the Greek thought that shaped and stayed in medieval thought, one must rise above the senses and bodily passions not merely to know God but to know this sensible reality as human, as a person.

By Thomas's epistemology, neither Mary in Bethlehem nor any other mother knows the humanity of her child from the bodily experience of Him at her breast. Nor do sexual lovers know each other by their bodily embrace. The senses cannot perceive distinctively human nature or any distinctively human traits as reason, will, and corresponding affections, for everything distinctively human is immaterial. Only by rising from physical experience to a higher, nonbodily kind of knowing, whether of faith or reason, can a mother attain her child as human. Playing with any human baby or holding any dying person misses the point, except inasmuch as one is spiritually aware of the baby's or man's spiritual soul.

If traditional Christian theology be right, no bodily experience itself can be knowing of another person, human or divine. I need to have the physical experience of the person in mind when I think of him, but my knowing the person is the thinking or rational understanding, not the physical experience. Only by my exercise of intelligence does the reality of the other person enter my awareness. Margaret of Cortona, however, and Mary of Oignies and Mechtild of Magdeburg knew that the physical experience they had of Christ was itself a knowing of him. In the very physical experience they were aware of the reality of Christ as a person.

It is similar with the Pietà and the devout onlooker. By traditional Christian philosophy, the essential conscious union of the onlooker with the mourning mother is immaterial, nonphysical. It consists in knowing Mary by bodiless intellect and loving her and sorrowing with her by bodiless will. The viewer's physical experience of the sculpture is necessary only as a context for the viewer to unite with Mary spiritually. The physical experience of the viewer also serves as an echo chamber where the intense union of intellect and will with Mary, the spiritual pain, may flow back into the viewer's body, causing bodily pain. Lack of this overflow, says Thomas Aquinas, justifies the suspicion that the spiritual union is deficient. But the bodily pain or sadness is a union with Mary mourning, a knowing of Mary mourning, only to the extent that the sadness reflects and expresses the spiritual sorrow of this individual viewer (ST I–II, 4, 6; II–II, 28, 3; *Compendium Theologiae*, 168, cited by Miles 1981, 132–34).

This traditional theory is contradicted by the impact of the Pietà on many devout believers. We can, I believe, imagine what must be such a pious experience, whether medieval or modern. The direct conscious union of the devout person with the mourning mother is inextricably both physical and personal. The union, which is, among other things, knowing the mourning mother, being aware of her, is wholly physical and wholly personal. The physical is the personal. The personal is the physical.

This is not a momentary confusion of consciousness. After the devout person's experience of the Pietà, she cannot go back in memory and distinguish in the original experience a nonbodily union with Mary from a bodily one. If the devout person were to abstract from his or her bodily aching with the griefstruck Mary, there would be little of moment left of that first, immediate experience of Mary. The aching of grief is a different kind of physical ache from the aching, say, of some illness or after extraordinarily strenuous bodily exercise. But the aching of grief *is* physical.

Going back in memory, such devout onlookers cannot extract or even point to a bodiless core or height or depth to the experience. They are with Mary *inasmuch as* they ache with Mary's aching. Apart from the ache they are not with her. Apart from the ache they are scarcely aware of her. The ache is their knowing and loving of her. The ache is their being with her. It is personal *qua* muscular, muscular *qua* personal. For our question about knowing, we can say this is knowing *qua* muscular, muscular *qua* knowing.

This knowing which pious medieval women gain and treasure challenges not only medieval establishment thinkers. It challenges, I submit, establishment intellectuals of the waning twentieth century. Take, for example, a college professor teaching on the morality of war or racism. Does she adequately know, can she adequately teach, and can her students adequately come to know the evil of war or racism if both the professor and students do nothing but disembodied reflection on statistical data and rationally analyzed events? If one day a student were for the first time gripped by physical pain as he looks at a photograph of a starving, abandoned child, a victim of war, does he now not know something important that he did not know before? To what extent can articles in an ethics journal communicate genuine knowing of any good or evil if the articles evoke no visceral response of their readers?

I am recalling the ongoing philosophical question of this second part of the book. What might indeed characterize a "bodily" knowing of personal good and evil? The philosophical question intertwines with the following historical inquiry into the kind of bodily knowing claimed by medieval women.

The foundation of mainstream Christian thought up through the Middle Ages was that only one kind of knowing was intrinsically worthwhile. Only one kind of knowing could know anything of importance. It was called "rational" or "intellectual" or "immaterial" or "faith seeking *intellectum*." In women's devotional experience there emerged into light another kind or cluster of kinds of knowing which they prized for its own sake. For them it was a unique way of knowing something precious, something personal.

The other kind(s) of human knowing did not return to deep shadow in the following centuries, but the light on it in Western culture remained sparse and flickering and is still so today. Western thought pays mainly lip service to it now, for, whether secular or religious, Western thought still recognizes only one intrinsically worthwhile kind of knowledge. As Heidegger said, Western thought still speaks Greek. One can put it more accurately: Western thought still speaks like a Greek man. (It does not speak as Sappho did.) Such thought can do little with modern women's experience. Like Joanna Ziegler, I see the devout women of the late Middle Ages as among the first moderns (Ziegler 1985a, 9–10).

It is enlightening to examine, as I am about to do, the characteristics of the women's experience which repelled the theologians and which the women treasured. The theologians do not deny that there can be human knowledge with these traits. But such traits cannot characterize knowing of any intrinsic importance because, the theologians say, they characterize the lowest kind of human knowledge which at best serves higher forms. They characterize that "bodiliness" that the theologians could not accord to any intrinsically worthwhile knowing. No knowing of the distinctively human, the personal, can be bodily.

Much bodily knowing is, of course, of little intrinsic value or importance and is at best useful because it serves higher knowledge. Such would be an itch, a feeling of having a full stomach, or sleepiness, when perhaps through other mental absorption the individual is aware of nothing but the itch, and so on. The bodily knowing that the devout women prized has some of the traits of the mere itch, full stomach, and sleepiness, but it has also other peculiar characteristics. These characteristics form a particular constellation, one kind, or cluster of kinds, of bodily knowing. It is on this particular constellation that our questioning converges: "What are these characteristics? And what exactly does the individual know by this knowing that is not other-wise known?"

Each of the characteristics of this kind of pious experience is the contrary of a characteristic of that distinctively human, intrinsically precious knowledge for which alone, according to the theologians, the human mind strives. The primary value which the theologians see in knowledge with the characteristics of the women's experience is extrinsic to it: it helps the mind to higher knowledge, knowledge with contrary characteristics, or it expresses or resonates with that higher knowledge. Only this higher knowledge with contrary characteristics to the women's experience is specifically human and knows the specif-ically human, the personal.

But though they might not be able to put it in clearcut abstract concepts, the women who felt bodily raptures or bodily devotion of this sort believed that they were being privileged to know in an intrin-sically valuable way, independently of any further knowledge that might result or prior knowledge that they had. They were aware of the reality known to an extent that no other knowledge enabled. The real-ity was person and personal. For both theologians and women the issue was a unique knowing of a person and all that made up their personal reality.

What then characterizes this kind of knowing claimed by the women and scorned by the theologians? Since I want to evoke the characteristics as they appear in consciousness, I may be permitted to start with my own experience. I recall summers with friends thirty years ago in Belgium, eastern France, and western Germany. Sitting in the back of the car, I directed the driver on the trail of a piece of Grunewald or Riemenschneider or an anonymous Pietà until I stood staring at it. The mosaics of Ravenna, the Christ of Chartres did not soak into me as this northern fourteenth- and fifteenth-century art did. Contemporary works of the south, say, of Giotto and Fra Angelico, whom I loved, did not move me half as much.

Then back into the car I went to pick up my *Michelin* guide and plot the route to the next nearest piece of northern fourteenth- or fif-teenth-century art. Why? What in this art drew me in as no other Christian art? I know now, with the clarity of retrospect. For one thing, this art brought me to rest. Before a Pietà I rested in pain. In every line of stone or wood, I felt Christ, God and man, dead for love of me. I felt the mother grieving him dead. I felt my own grief. That was all I felt. I felt nothing else at the moment, nor wanted to.

I felt Christ's love realized in his death. I experienced his love as divine as well as human, but the sculpture did not send me on to divine attributes of this love. No line of wood or stone lifted my mind to anything of God's salvific plan, his gracious condescension, his infinite wisdom or even the wonder that he chose from all eternity to become human and live and die for me.

In earlier art I had seen a Crucified Christ with the Resurrection already glowing on the pain-wracked face. As it pained me, it sent me on to Christ's eternal glory. Not so this Flemish and west German art of the late Middle Ages. It surrounded me, enclosed me. It closed me in a single moment of time at a single place in anguish with Christ and his mother. My experience did move me after a while to meditate and appreciate better some spiritual reality, like the infinity of God's love for me. But that was not the innate impact of the sculpture or painting. The art itself pinned me down within the Passion of Christ and Mary. It held me fast in their pain and sorrow.

Other was the impact of earlier Christian art which had also moved me. That art did not hold me in the physical experience of what it described. That art rose from the physical beyond the physical, carrying the viewer with it. Earlier Christian art was primarily symbolic or allegorical or didactic, as one sees in the sculpture abounding on the earlier cathedrals or the images of the illuminated Bibles.

Even the realistic art of the South I had seen drew the viewer into a Bible scene in such a way as to lift him or her to intimated nonmaterial realities. The glowing visage of Fra Angelico's Virgin sends us immediately inward and upward to a spiritual message she must be hearing. Giotto's apostles are transparently moved and absorbed by an event that only faith discerns. But the mother of the northern Pietà holding the dead Jesus holds our awed, sad feelings down with hers. She and her son are all we know. For the moment, they are all we want to know. So too, in similar art, say, a woodcarved altar piece by Riemenschneider, when the mother bends tenderly over the newborn child, she holds our awed, glad feelings with her. She and her son are all we know and all we want to know.

Joanna Ziegler pointed out on slide projections for me and others how the lines of the Pietàs of northern Europe (unlike Michelangelo's later one) lead the eye inward, binding the two figures together. In the slides we watched, the inward interlocking became stronger as the date of the medieval piece was later. She pointed out, too, how this is clearly not a timeless present but a moment of time, stained by past and future, by memory and expectation Past and future are here in the present: both the agony of the cross and the coming Resurrection. But they are piled into the present moment.

I suggest that the mother does not see the agony just past. But her muscles remember in their ache and weariness. She does not foresee in her mind's eye a resurrection, but her muscles in a subliminal steadiness beneath the present grief know something like a resurrection will come. But for Mary, the past and future at the moment are dim at the edges. Dull grief at the limp body in her lap holds the center.

The satisfying knowing which a devout Christian had of a sculpted Pietà or other northern late medieval art was for traditional Christian theology knowledge of the lowest kind, of minimal value. One reason was that it was knowledge of the singular, i.e., simply of individual realities as individual in the reality that that individual was at an individual moment in an individual place. The mind came to rest there, held fast, and was held fast. Truly worthwhile knowledge, the theologians knew, is always knowledge of the universal. It is knowing of human individuals precisely by attributes that they had in common with others (such as rationality or substance or joy or love). This knowing moved ever onward toward knowing God better as the totality of reality, of which all individuals were only participations (ST I, 84–86). God's reality is infinitely more known in knowing that he is Being Itself than in knowing that he is an existing individual. Angels were both universal and individual. Their nature was their individuality. Each individual angel exhausted its species.

How then could there be individual human beings when their common human nature did nothing to individuate them? A frequent answer by the theologians, an answer voiced first by Aristotle, was that only the individual's connection to the body made possible that he or she was this human person and not that one. The distinctness of one human being from another was a limitation, an inferiority, in comparison with higher beings.

By traditional Christian theology, the individual human being is indeed of great worth. Indeed, humans are of infinite worth, for not only are they the image of God but God loves them and wills that they be with him for all eternity. But to know and fully appreciate their worth, we need know only the nature and relation to God that all human beings have. It adds nothing to know a person in his or her singular existence and life.

The traditional Western optic is exemplified by Thomas's natural law ethics of war. What makes a war just or unjust is a weighing of its consequences on the citizens *qua* sharing in the common good and *capaces beatitudinis*. There is not the slightest need to open one's mind to the reality of any individual citizen undergoing the destruction of war. Those who think this is a problem peculiar to medieval moral theology might recall the discourse of authors, politicians, and professors before the recent Persian Gulf War. Rational generalities are essential. The issue is: Do they give us the whole pertinent reality of war? The question is pertinent whether or not the final judgment on the justice and wisdom of the war would be changed by so opening one's mind to this or that singular individual, real or imaginary, in the agony and loss of war victims.

The medieval art we are considering, such as the Pietà, locked the responding Christian into a single time and place with individual human beings. It closed them in with Jesus and Mary at a given moment of their lives. The experience the art evoked did not contain and was not in motion toward more spiritual truth. The experience did not contain nor was it in motion toward *any* universal truth. It was not even in motion toward that moment two days later when Jesus would rise again, though neither is it despair without intimation of such eventuality. The art buried the Christian in a moment of time, in a single moment of single intersection of individuals' lives. Wherefore it was for the theologians least knowing and for the women's great knowing.

But often enough in this kind of experience, medieval women, unlike myself decades ago, were far from passive observers. The experience has often a second trait, which I missed completely in the earlier published version of these pages. Students of mine pointed it out to me. The women were active. Kneeling or standing before the Pietà they identified with Mary and with Jesus whom she held. In many of the Pietàs, though no movement is portrayed, the arms and body of the mother support with firmness or encircle with tenderness the limp son. Similarly active in devout fantasy was Margaret of Oingt, who wrote:

> When the day of the Nativity of Jesus Christ came, I spiritually took the glorious child into my arms. Thus I carried and kissed Him tenderly in the arms of my heart, from the hour of matins until tierce....

> At noon, I reflected how my sweet Lord was tormented for our sins and suspended all naked from the cross between two thieves. When I came to the point where the evil people had deserted Him, I went toward Him, with great respect, and removed the nails; then I took Him upon my shoulders and took Him down from the cross and put Him in the arms of my heart, and it seemed to me that I carried Him as easily as if He had been one year old....

> In the evening when I was going to lie down, in my spirit I put Him down on my bed, and I kissed His tender hands and these

blessed feet which were so cruelly pierced for our sins. And then I bent down toward that glorious flank which was so cruelly wounded for me. (1990, 63)

An awareness of the unprecedented activeness of women in their devout union with Christ led me to recognize in much of this experience the unprecedented mutuality on which Part I of this book concentrates. For a modern reader who is familiar with certain contemporary Christian piety, as illustrated by a crèche or Crucifix, the women's pious agency, their affecting Christ as he affects them, may seem unremarkable, perhaps even sentimental and pathetic. But there was rarely anything like it in the first millennium of Christianity. What is new here? Whether or not its expression be banal, sentimental, or pathetic, this is an emergence of a new ideal for knowing as well as for interpersonal relating generally. We scrutinize these beginnings because we ask whether they reveal an evolutionary change not merely in prized religious experience but in prized human experience, religious or secular.

One who is aware of Christian piety of the first millennium must marvel at the growing agency of devout Christians, especially women, as the second millennium begins and progresses. One sees it over and over in devotions such as those Margaret of Oingt recounts above. One sees it over and over in art like the Pietà, with which the devout identified. One sees it in myriad accounts of mystical experience or "visions," such as Hadewijch's, as we brought out in Part I.

This, as we noted a propos of Hadewijch, was another kind of loving from what the theologians recognized. But it was also another kind of knowing. Hadewijch stated more than once that her agency and affecting of Christ, yielded her "perfect knowing" of God. Her "conquest" of Divine Love drew Love's embrace of her, a single individual at a single moment of time and place. Through their embrace she knew him as never before. By "knowing" I mean simply the presence in an individual's awareness of another reality. In her mutuality, her interaffecting with God, their flowing back and forth into each other, drinking each other's love, drawing and conquering, satisfying and contenting each other, Hadewijch knew God in a unique, supreme way. How is it a knowing? Consider further characteristics common to Hadewijch's supreme experience and that of many other devout women of her age.

# 10

Many devout experiences, at times ecstatic, as reported by the women and as reflected in the popular art, not only held the individual believer to knowing individual persons at a single time and place in their life. Not only was the devout woman active in that time and place with Christ or other individual persons. The same experiences have often a third characteristic that is largely absent in descriptions of religious experience by the theologians. Bodily touching pervades much of the treasured experience of the women. The religious experience of medieval women which we have labeled as "bodily" is often tactile to an extent unprecedented in the first millennium of Christian experience. By "tactile" and "touch" I mean what the establishment thinkers of the Western philosophical and theological tradition meant by it. We moderns might not use "touch" to designate all these sensations, but I believe we can see their similarity to each other and their difference from the four other external senses: seeing, hearing, taste, and smell.

Recall Low Countries sculpture, as exemplified by the Holy Cross exhibition. Ziegler spoke of the "physical and psychological states of being" of Jesus, Mary, and followers that the viewers are brought to share. The viewers first see a "twisted grimace beneath the thorns, sagging shoulders beneath the cross, open mouth of horror." But then, by empathy, even a secular, modern viewer is likely to feel the pain from the piercing thorns, the crushing weight of the cross, the gasping for breath. Which of the five external senses are involved here? The sense of touch. That is the sense by which we become aware of thorns piercing and cross crushing and our gasping. So too, by touch are Mary and we aware of the dead Son held in our lap. By touch are Mary Magdalen and we aware of the foot of the cross which we embrace with arms and thighs.

Recall the widespread trend of the time for women to desire fiercely to eat the sacred Bread and drink the Wine. The taste was not important for what they desired, though some reported a wonderful honeylike taste. Not only looking at bread and wine up on the altar, but the taking them in the mouth, the swallowing it, was essential. These are experiences of the sense of touch, at least as the male thinkers of the West had categorized that sense. The male category serves pragmatically our inquiry, for our goal is to identify as best we can the kinds of knowing which the theologians denigrated and the devout faithful prized.

As Bynum notes, in the eucharistic raptures of some women, the wounds of Christ appeared spontaneously in their hands, feet, sides, and faces. Women who did not experience such mystical or paramystical phenomena felt similarly, though in a more moderate way. "No religious woman failed to see and feel Jesus as wounded, bleeding and dying" (1984,189). But feeling of wounds had to be tactile, not merely visual. Bynum thus confirms Ziegler's inference from sculpture such as the Pietà: women desired in prayer to take the dead Jesus down from the cross onto their laps.

Recall, too that the physical union the women sought and had with Christ was not only that of sharing his agony and passion. They saw often in the Host and chalice Christ the baby or Christ the bridegroom and joined him as such. "Agnes of Montepulciano and Margaret of Faenza became so intoxicated with the pleasure of holding the baby that they refused to give him up. Ida of Louvain bathed and played with him. Gertrude of Helfta, Lidwina of Schiedam, and Gertrude of Delft nursed him at their breasts." (Bynum 1985a, 9–10; cf. 1984, 188–89).

Listen again to Margaret of Oingt:

> When the day of the Nativity of Jesus Christ came, I spiritually took the glorious child into my arms. Thus I carried and kissed Him tenderly in the arms of my heart, from the hour of matins until tierce....

> At noon, I reflected how my sweet Lord was tormented for our sins and suspended all naked from the cross between two thieves. When I came to the point where the evil people had deserted Him, I went toward Him, with great respect, and removed the nails; then I took Him upon my shoulders and took Him down from the cross and put Him in the arms of my heart; and it seemed to me that I carried Him as easily as if He had been one year old....

> In the evening when I was going to lie down, in my spirit I put Him down on my bed and I kissed His tender hands and these blessed feet which were so cruelly pierced for our sins. And then

I bent down toward that glorious flank which was so cruelly pierced for our sins. And there I commended myself and my brother to Him and asked forgiveness for our sins; and thus I rested until matins. (1990, 63)

Blumenfeld-Kosinski comments here in a footnote on Margaret's phrase "arms of her heart": "That is, not her physical arms. Although the experience seems to be described in physical terms, it is actually purely spiritual" (63). Margaret, too, says she "spiritually took the child in her arms." Blumenfeld-Kosinski refers in her interpretive essay to a vision of Margaret representing "a flowering and consequently a validation of the five corporeal senses." Blumenfeld-Kosinski seems to say here that the senses' "importance in experiencing God" is always metaphorical except for the use of senses in reading a text: "The experiences which could be transmitted directly through the senses thus seem to be mediated by writing."

I wonder whether Blumenfeld-Kosinski is unduly influenced by both medieval and late twentieth-century philosophical concepts. It is impossible, I submit, to read the lines of Margaret quoted above, and believe she is merely using subsequently imagined metaphors to describe what was originally a purely intellectual or imageless experience. She reports a sensual experience, which she then interprets according to establishment theology.

Some modern readers might judge the experience to be "illusion" or "fantasy" or "mere imagination." No one, surely, would judge it to be imageless or nonsensory or mainly intellectual. Margaret knew well that it was not ordinary sensual experience of the body and probably believed, with the ambient theology, that her own body played no part in it. It suffices for my point that she experienced the experience as of a bodily kind. For example, she believed she had truly held Christ's body.

We do not find in Margaret's text any mediate step from her imagery to reality known. We do not even find— *pace* Blumenfeld-Kosinski—a transforming step from experience to word. Margaret reports directly what happened to her. Contrast her account with the explanation, discussed in Part I, of his erotic imagery which pervades the text of Bernard of Clairvaux when he tells monks to unite with God by kiss and embrace. Bernard believes he is using a comparison to instruct his hearers and readers. Margaret tells us directly what happened to her and what she did.

On the face of it, this extensive, intensive use of touch, even as mere metaphor or image, in describing experience of the divine is foreign to the established Western tradition up through Plotinus. Plato,

Plotinus, and the tradition describe the human individual's knowing the divine as an ascent whose summit has two stages. The second is even higher, nobler, and more completely satisfying than the first. In the prior stage, the individual "sees" the divine in different, distinct aspects. In the second stage, the individual comes to know the divine more profoundly and in its unity rather than in plurality. The knowing of the second stage, i.e., the knowing of Beauty Itself or Goodness Itself, is for Plato and Plotinus something like a bodily sensation. It is like the bodily sensation of seeing.

The stage of the ascent in Plato's *Symposium* immediately before the final goal is that of the "branches of knowledge." Here the human being will see also their beauty and

> by looking thus on beauty in the mass may escape from the mean, meticulous slavery of a single instance where he must centre his care like a lackey, upon the beauty of a particular child or man or single observance; and turning rather toward the main ocean of the beautiful may by the contemplation of this bring forth in all their splendour many fair fruits of discourse and meditation in a plenteous crop of philosophy.

The summit of the ascent, however, is to suddenly "see" something wonderful, beautiful by nature, that for the sake of which were all the previous toils. It is eternal, neither waxing nor waning. It is beautiful at every time in every part and aspect. It exists in itself with itself. It is beauty itself that the human being now gazes on. It is the single form of beauty in which all other beautiful things share. "There in life, dear Socrates, above all other places, said the Mantinean woman, should a human being live, contemplating beauty itself" (*Symposium*, 210E–212A; my translation).

Whether Plato had a mystical experience is not clear. Porphyry reports that Plotinus did. Plotinus founded Neoplatonism, the main thought that moved traditional Western thought along and provided a framework for mystical theologians, male as well as female. It is a frame the women accept but also use to advance further. They use the framework to break out of it and build alongside it and, in part, on it. For the moment, we concern ourselves simply with the issue of vision versus touch in the supreme union of a human being with God. Plotinus speaks often of this supreme union. He describes it as precisely that: union, perfect, complete union of the individual with the Divine. That is the truest one can say of it: it is becoming one with the One. It is a union for which the individual after the experience has no word or idea to express the union.

Nevertheless, Plotinus makes some effort to identify further what this union is. He describes it at times as a kind of "touching" (*thigein*, e.g., VI, 9, 4, 27; 7, 4), "laying hold of" (*ephaptesthai* or *epaphe*, e.g., VI, 9, 4, 27; 9, 56; 7, 25; 9, 19). Much more often he speaks of this union as, though different from the more intelligible "seeing" of the prior stage, still itself a kind of "seeing" (*horan* or *idein*, e.g., VI, 9, 4, 16, 29; 9, 1), "contemplation" (*theaomai* or *thea*, in the etymological sense of "gaze on, behold," e.g., VI, 9, 4, 14, 30; 7, 20).

> It is possible here to see [*horan*] the One and oneself, inasmuch as it is permitted to see. One sees oneself bursting with light and filled with intelligible light. Rather one becomes oneself pure light, light, weightless. One becomes, rather one is god. (VI, 9, 9, 56–59; my translation based on Bréhier's)

> It is not afraid of any misfortune while it is with This and while it has the full vision.

> …but requiring real contemplation, he would…look only at him in future; and then, looking at him and not taking his eyes off him, by the continuity of his gaze he would no longer see a sight but blend his vision with its object, so that what he saw before became sight in him.

> The soul sees Him by a kind of blurring together the *Nous* abiding in it and making it disappear, or rather its *Nous* sees first, and the contemplation passes to it and the two become one. The Good is spread out upon them and united with the combination of both, and runs over the two and rests upon them, uniting them and giving them a blessed sense and sight [*aisthesis kai thean*].

> Then…led by his instruction to *Nous* and firmly established in beauty, he raises his thought to that in which he is, but is carried out of it by the very surge of the wave of *Nous* and lifted high by its swell, suddenly sees without knowing how; the Sight fills his eyes with light but does not make him see something else by it, but the Light is That Which he sees. There is not in It one thing which is seen and another which is Its light, or *Nous* and that which it thinks. (from VI, 7, 34–36; Armstrong's translation)

> This is the soul's true end, to touch that Light and see It by Itself, not by another light, by Itself, Which gives it sight as well. It must see That Light by which it is enlightened; for we do not see the sun by another light than his own. (V, 3, 17; Armstrong's translation)

That the supreme union of a person with God is a transcendent kind of seeing is echoed in subsequent Christian thinkers, such as Pseudo-Dionysius. By the thirteenth century, Thomas Aquinas, the establishment theologian destined to influence most the continuation of medieval theology into the modern period, will still identify this supreme union as a kind of seeing. Though careful to avoid pantheistic overtones, he will repeat the visual language of Plotinus and Pseudo-Dionysius.

True, more and more Christian thinkers in the thousand years after Plotinus also echo his passing assertion that the supreme union of a person with God can be compared with a kind of touching. They echo, too, Plotinus's passing suggestion of a similarity between the soul's union and the union of human lovers (VI, 9, 29–42). This suggestion is orchestrated by the commentaries on the Song of Songs by Origen and his successors, including preeminently twelfth-century theologians such as Bernard of Clairvaux. But to my knowledge, Hadewijch, writing in the first half of the thirteenth century, is the first systematic theologian to identify squarely and present systematically this supreme union as itself a tactile kind of experience. Both Bernard and she call consistently this union an embrace. But only Bernard, not Hadewijch, goes on, as we saw in Part I, to interpret the embrace in nontactile terms, i.e., not only explicitly as spiritual but also as an identification with the Word for which touching cannot serve even as metaphor.

Thomas Aquinas also ignores the humanistic import of traditional erotic metaphors. These thinkers have to ignore it because, as we saw, their epistemology excludes the possibility of any knowledge that would be intrinsically bodily and yet knowledge of a human person as such. Sense knowledge is useful and necessary for human beings but is never itself knowledge of the distinctively human. This is peculiarly true of knowing by bodily touch, such as in sexual union. Seeing beautiful things can contribute to and share enough in the intellectual vision it yields that bodily seeing has a certain worth for itself. But not bodily touching, whether in sex, eating, drinking or whatever. Why?

In the hierarchy of knowledge, bodily knowledge is lowest. Bodily knowledge, says Thomas, is least knowledge, hardly knowledge at all, because it is sense knowledge. Sense knowledge is wholly inferior to the rest of human knowledge, which is rational knowledge. Only reason can know being and its properties: goodness, beauty, and unity. The senses know nothing of this. Humans can perceive by eye and ear the beauty of sound and sight only by these senses' participation in reason.

Sense knowledge itself is of some worth inasmuch as it is knowledge and knows its object. Thomas rates the different senses as more or less of "worth" (*bonus*) inasmuch as each is more or less knowing. The sense lowest in value is the sense that knows the least. It is the sense of touch, says Thomas, following Aristotle. The senses rank in value to the degree in which they resemble reason. The sense of touch is least like reason.

This is not to deny that touch, as Margaret Miles brings out, for Thomas, following Aristotle, is the only sense in which humans surpass the animals, since the former's touch is finer and more assured. Touch is the basis of all the other senses, and excellence of reason requires corresponding excellence of sense. People have greater ingenuity and excellence of mind who have delicate touch. (Miles cites Thomas Aquinas's *Commentary on Aristotle*, II de Anima, lect. 19, and S.th. I, 76, 5.) But none of this contradicts Thomas's assertion elsewhere that a person's touch, compared with the other senses and especially to reason, is least knowing. Moreover, the kinds of touch commonly felt in the women's devotion, e.g., in satisfying hunger or sex appetite or in hugging or in weariness, are not what Thomas would call "finer" or "delicate," indicating ingenuity or excellence of mind.

That human touch is least like human reason and thus least knowing is why, for Thomas Aquinas, it is a sin to choose to eat or drink or have sex for the pleasure of eating, drinking, or having sex. There is nothing wrong in principle in choosing a reality for the pleasure it gives. All of us, if we are wise, choose God for the pleasure of union with Him. It is not a sin, but good indeed, to choose to get to know a friend for the pleasure of doing so. It is not a sin, but good, to choose for its own sake the enjoyment of seeing visible beauty or hearing beautiful music. But it is a sin to choose to eat, drink or have sex for the pleasure of it. Why the difference? The pleasures of eating, drinking, or having sex are pleasures of touch.

The pleasures of touch are good, says Thomas Aquinas, for created human nature remains essentially good even after the Fall. But the pleasures of touch are the least good of all human pleasures because the perceptions we take pleasure in with touch are the minimal forms of knowing. The sense of touch never participates in reason, though higher senses do on occasion. For this reason, pleasures of sight and sound can be worth being chosen for their own sake. Never should the pleasures of touch be so chosen, for touch is minimal knowing, nonrational knowing. We may not, therefore, choose to eat, drink, and have sex for their pleasures. We may choose them only for further good they effect, e.g., self-preservation, nourishment, continuance of the species, and avoidance of greater sin. In the Appendix, I document and detail further this reasoning of Thomas. I also point to clues in his text of what implicit value judgments determined Thomas's narrow and rigid conception of the only knowledge that counts, which he calls "reason."

But bodily touching suffuses the pious and mystical experiences we are studying. The women seek and enjoy eating and drinking Christ, caressing and nursing Christ, making love and playing with Christ, embracing the suffering or dead Christ. All are kinds of bodily touching. The women declare, contradicting unwittingly Thomas Aquinas and medieval theologians generally, that their tactile experience is a sublime knowing of Christ, a wondrous experience of Him worth seeking for its own sake, without reference to any further good achieved or achievable.

Other writers, including modern historians, betray perhaps the continuing influence of traditional Western thought when they refer to these experiences as "visions." Vision is theologically the most respectable and honored of the five senses. But in many of the religious experiences we consider, vision is subordinated to touch. The rapture, ecstacy, knowing of Christ is in the touching, holding. Hadewijch, as we saw, exemplifies this forcefully.

Hadewijch, like Plato and Plotinus, distinguishes two stages in the final supreme experience that the soul can have of God. For the prior stage, she follows the tradition in stressing "seeing" and "hearing," though she brings in more dramatically and centrally the physically visual and auditory than did her theological contemporaries and predecessors (e.g., Vision 6). But in describing the second, most ultimate stage, Hadewijch, like the devout medieval women we have been considering, uses "seeing" rarely and almost always "touch" to designate this final union. Noteworthy, too, is Plotinus's and Bernard's metaphor that the soul is embraced as it strives upward in ascent and in the embrace remains on the summit. Hadewijch, having made the ascent to the seeing level, now on that level is embraced by *and embraces* Divine Love and then falls down into an abyss.

> But then wonder seized me because of all the riches I had seen in him, and through this wonder I came out of the spirit in which I had seen all that I sought; and as in this situation in all this rich enlightenment I recognized my awe-inspiring, my unspeakably sweet Beloved, I fell out of the spirit—from myself and all I had seen in him—and, wholly lost, fell upon the breast, the fruition of his Nature, which is Love. There I remained, engulfed and lost, without any comprehension of other knowledge, or sight, or spiritual understanding, except to be one with him and to have fruition of this union. (Vision 6, 76f.)

After that [her reception of the Eucharist], he came himself to me, took me entirely in his arms, and pressed me to him, and all my members felt his in full felicity, in accordance with the desire of my heart and my humanity. So I was outwardly satisfied and fully transported. Also then, for a short while, I had the strength to bear this; but soon, after a short time, I lost that manly beauty outwardly in the sight of his form. I saw him completely come to nought and so fade and all at once dissolve that I could no longer distinguish him within me. Then it was to me as if we were one without difference....After that I remained in a passing away in my beloved, so that I wholly melted in him and nothing any longer remained to me of myself. (Vision 7, 280–82)

But Love came and embraced me, and I came out of the spirit and remained lying until late in the day, inebriated [verdronken] with unspeakable wonders. (Vision 9, 286f.)

The Voice embraced me with an unheard of wonder, and I swooned in it, and my spirit failed me to see or hear more. And I lay in this fruition half an hour; but then the night was over, and I came back, piteously lamenting my exile, as I have done all this winter. (Vision 10, 288f.)

In that abyss I saw myself swallowed up. Then I received the certainty of being received, in this form, in my Beloved, and my Beloved also in me. (Vision 12, 296f.)

Then fruition overcame me as before and I sank into the fathomless depth and came out of the spirit in that hour, of which one can never speak at all. (Vision 13, 302f.)

Once I lay for three days and for the same number of nights in entrancement of Spirit at the Countenance of our Beloved; and this has often lasted for that length of time; and also for the same length of time entirely out of the spirit, lost here to myself and all persons, in fruition of him: to know how in fruition he embraces himself. To be out of the spirit and to be in him—this surpasses all that one can have from him and all that he himself can accomplish, and then one is not less himself than he is. (Vision 14, 305f.)

Hadewijch does not record that she felt Christ as hot or cold, rough or smooth, hard or soft. Rather, as we have just heard, she felt herself embraced by him (Vision 6, 7, 9, 10, and 14), taken in his arms (7), pressed by him to his body so all her members feel his (7). She felt herself lying on the breast of his Nature of Love (6 and 7). In her fruition, seeing and hearing no more (10), she felt herself engulfed (6), swallowed up in the abyss (12), sinking into this fathomless depth (13). If in your epistemology there are only five external senses, then these sensations of Hadewijch are all those of touch. However you classify them, these sense experiences constitute a group that Western thinkers never applied to divine union as intensively and systematically as Hadewijch.

In that Hadewijch's supreme experience of God is of a touching kind, it is one with other religious experiences of medieval women referred to above. Should we be surprised that for medieval women a profound, unique, precious knowing takes place in something much more like touch than seeing? Does not some of the greatest awareness in family take place through touch? Is it not at times the most precious mutual awareness of each other that the family has? Tactile sensations not only involve eating and drinking together, giving suck, playing with the baby, and having sex, which are described in the women's "visions." (For traditional thought, touch is the sensation involved in orgasm, swallowing, and sating hunger and thirst.) But touch is also central to supporting the baby taking its first steps, clothing the child, mutely embracing the bereaved, caring for the sick, holding the ill person's hand, sleeping together, the playful wrestling of children, or the adult hug of reunion after long separation. This contradicts, as we will see shortly, traditional medieval thinkers such as Thomas Aquinas. Are we not in these experiences "more knowing" than usual? The depth and extent of tactile knowing can also make an act more evil, as in a slap or kick or rape.

Let me, less wise, make a suggestion to the art historians. Joanna Ziegler asks, "Why did the Pietà originate—and in essence remain—a *sculptural, rather than painted,* type?" (1985b, 13) Could it be because devout women found special satisfaction in identifying with the sorrowing mother in the feelings of touch throughout her body, her tactile holding, weariness, pain, and abandon? Does not sculpture by its three-dimensionality convey more powerfully than painting a tactile experience? The wood or stone filling space draws us out through our eye to feel her sad, limp, tired pain through head and shoulders and entire body?

In the last part of her Brown University lecture, "The Emergence of a Women's Sensibility in Late Medieval Art in Northern Europe," Ziegler began to answer her question. She showed in detail how sculpture, compared with other art forms such as painting, "was ideally suited to visualizing for the laity the humanness, the bodiliness, the physicality of Christ and Mary" (1986b, 16). She did not discuss, however, whether sculpture did this because in part it ideally communicated in the visualizing the tactile experience of Christ and Mary.

## 11

The physical experiences that the medieval women had of Christ have often a fourth trait that was depreciated generally by traditional Christian thinkers. The women's physical experience of Christ is at times the kind of physical experience we have when we happen to perceive something and are moved by the perception, e.g., a lovely landscape, a moving piece of music or work of art, or a beautiful person. Such experience was not depreciated, but extolled by traditional thinkers. The object is perceived by us and we give our attention to it. By our very perception and attention we know its beauty and goodness. As a result of our knowing, our contemplation, we cannot help but love it: enjoy it and desire to continue to contemplate it.

The object may be something we have known and loved before. We repeat the process on this new encounter with it. As I noted with respect to the sense of touch, this kind of physical experience through seeing, hearing, and smelling the theologians ranked high among forms of bodily knowledge, though it is only by reason entering into the experience that the individual perceives the beauty or goodness as such.

Often, however, the women do not have this contemplative experience, but—and this is the fourth trait of knowing looked down on by the establishment thinkers—an experience of a contrary kind. It is an experience we humans have only after and because and inasmuch as desire rises in us. The now pressing need yields us a kind of knowing that we do not have except when we feel this need. It is not enough for the object to be perceived and our attention given to it. The need or desire also must arise in us. Sometimes it does, sometimes not.

In ordinary human life, hunger, thirst, sleepiness and sexual desire, again, exemplify this. Feeling these desires makes us know something about what or whom we want and about ourselves. At no other time do we know it, except by remembering the times we felt this way. This kind of desire often does not move us serenely as the beauty of landscape or song. It presses us, urges us. It can disturb us, often violently. Unsatisfied, it can overwhelm us, make us wild.

The erotic mysticism of women like Hadewijch and Beatrice of Nazareth, Caroline Bynum tells us, is not that of passive, submissive brides. "Their search for Christ took them through a frenzy which they called insanity." So too, Mary of Oignies, "When she was not able to bear any longer her thirst for the vivifying blood...would remain for a long time contemplating the empty chalice on the altar." As we saw, a recurrent conflict in the church life of these times was the women demanding more frequent communion and the theologians resisting this. It was in the "heat" (*aestus*) of passion that the women experienced Christ (Bynum 1984, 192, 179). Hadewijch writes a number of times of supreme human love of God as when one "loves so violently that he fears he will lose his mind, and his heart feels oppression, and his veins continually stretch and rupture, and his soul melts" (Letter 8, 27ff.; cf., e.g., Vision 7, 1ff.). The women's need in these instances was thoroughly physical and bodily, as well as religious and personal. Yearning with desire, the women knew in a peculiar way him whom they desired.

That some knowledge can be had when and only when the individual feels certain desire is a commonplace of Western thought. The distinctive wisdom of the Christian, says Thomas Aquinas, is formed by the Christian's charity, the Christian's love of God. More generally, Christian theology, in the Middle Ages as today, echoes paeans by Plato and Augustine to the desire of the soul for the good and the beautiful (*Phaedrus*, 249–57; *Confessions*, IX, 10). Only this desire for the good and beautiful can open us to know the good and beautiful. Only to the extent that our desire becomes more conscious and we engage our whole self in it, does our rational or intellectual knowledge grow of what is and is good. Thomas follows Aristotle here, as do contemporary ethicists (ST I–II, 39–40; Gilson 1929, 347–49; Milhaven 1971, 421–30).

By this traditional epistemology of the West an individual's desire enables him to know because it moves and thus enables him to exercise his knowing power, to concentrate his attention and thus to see what is being disclosed to him. In a dramatic turn of the tradition, some Neoplatonic Christian thinkers changed the place of love and intellect in supreme union with God. Whereas some Christian thinkers, like Thomas Aquinas saw still the mystic's union with God as intellectual vision (ST II–II, 175), other theologians, like Thomas Gallus and the author of *The Cloud of Unknowing*, saw the supreme union of the mystic with God to be an affective awareness of surging, climaxing, and desiring love in darkness, bereft of corresponding knowing except by a rare, revealing "beam of light" (Hodgson 1944, lxii–lxiv, lxix). For Bernard of Clairvaux, too, and other twelfth-century theologians, the mystical union was of love rather than of intellect, though they did not favor images of intellectual darkness.

These mystical and contemplative theologians, later ones like the earlier, meant incorporeal love and desire. They did not mean bodily desire such as these devout women felt. The knowing which the theologians extolled depended in no way on a bodily need asserting itself. It came out of desire rising out of bodiless will and/or intellect acted on by God's grace. A traditional thinker could not write, meaning literally, what Audre Lorde writes in her poem, "On a Night of the Full Moon" (1976):

> Out of my flesh that hungers
> and my mouth that knows
> comes the shape I am seeking
> for reason.

The religious experience of women of the late Middle Ages often had a fifth characteristic that was in principle denigrated by theologians. This overlaps other characteristics I have listed so far, but its overlapping, I submit, enlightens. It reveals further the richness of the women's experience and the peculiarity of the belittling by the theologians.

Consider an argument that Thomas Aquinas uses to support the traditional prohibition of marriage between close relatives. The pleasure of sexual intercourse, he quotes Aristotle approvingly, corrupts seriously the judgment of moral wisdom. This pleasure, Thomas notes, is increased by the love of the persons in the union. If, therefore, he reasons, to the sexual love itself is added the love "that comes from common origin and sustenance," the soul "would be necessarily more overcome by the pleasure," and therefore its wisdom more corrupted (SCG III, 125; cf. ST II–II, 154, 9; Aristotle, VI: 5, 1140$^b$ 10–20).

Thomas and other theologians were ignorant of the wisdom, the "knowing," that grows out of sexual love and pleasure. More to our present point: they were ignorant of the wisdom coming out of the ordinary love and pleasure of "common origin and sustenance," of living together in family, of being engaged with and spending time with the same parents, grandparents, and other relatives, of eating and drinking day after day at the same table, as well as of making love with one's spouse. As we saw, the theologians did not ignore all knowledge arising from bodily pleasure. They praised the eyes' and ears' pleasurable perception of beauty, and the higher knowledge it enabled. They ignored the wisdom, the knowledge of value, arising from pleasure in satisfying basic human needs, the needs for sexual intercourse, life together, shelter, sustenance, and so on. They ignored, in general the wisdom of ordinary family life.

Many medieval women, like the theologians, knew that both sexual and other loving that can come with family life bring their special pleasures. They knew, unlike the theologians, that these pleasures constitute a personal union of the persons involved. In the very satisfaction, the everyday satisfaction, of elemental human needs, one experiences uniquely, comes to know profoundly the other person.

Here again, as in each of the four points I already made, the theologians failed to recognize in their conceptual syntheses what they with the rest of Christendom presupposed by their metaphors. Traditional thinkers describe regularly the life of the blest after death as a banquet, as happy family life. They emphasize how much the Eucharist is like a meal together. They describe often mystical union or Christ's union with the Church as sexual union. The Song of Songs was "the book which was most read, and most frequently commented in the medieval cloister" (Leclercq 1960, 106). The pleasures of meals, sex, and other family activities serve the tradition as images of, among other things, richly aware interpersonal union.

But traditional Christian thinkers failed to endorse explicitly what they implied in their metaphors.

> For monastic and secular literature on the whole, we can apply with some reservations this observation made by Edmond Faral, a historian as competent as he is impartial: "It is remarkable that the Song of Songs…has never been interpreted, in the many commentaries written from the tenth to the fourteenth century, otherwise than in the sense of religious sentiments and mysticism. No lay author ever saw in it, for his inspiration, any signs of human passion." (Leclercq 1979, 29; cf. 1960, 106–9)

Traditional Christian thinkers did not acknowledge that if they used a reality of this world to image something otherworldly, the worldly reality must be similar to the otherworldly reality. The worldly reality must have, if in a much lesser degree and a different way, those good qualities that the thinkers want by the metaphor to bring out in the otherworldly reality. Otherwise, the comparison is meaningless and groundless.

Bernard of Clairvaux and thinkers influenced by him went beyond previous Christian theology in expressing high esteem for married love, including its "carnal love" and "fleshly pleasure." But they never say that the carnal or fleshly dimension itself is a genuine knowing of the other, nor that it has any intrinsic value of its own distinct from the spiritual. When speaking of marital sex, they appear to ignore familiar biblical language where "knowing" is a synonym for sexual union (e.g., Gen. 24:16 and Luke 1:34). Leclercq in his *Monks on Marriage: A Twelfth-Century View* (1982) does not seem to note the monks' curious lapse.

In composing originally this essay on bodily knowing, as published in the *Journal of the Academy of Religion* (1989c), I went no further than this trait of the women's knowing, i.e., its likeness to familial pleasuring. In so doing, I illustrated unintentionally the resistance of the male intelligence (mine) to taking in this kind of experience. I said nothing of how this pleasuring, in everyday reality and in the women's devotion, is a "mutuality," in the sense I develop in Part I. I made the passing assertion that this familial pleasuring can "constitute a personal union of the persons involved" (363). I noted: "The pleasures of banquet, sex, and other family life serve the tradition as images of, among other things, richly aware interpersonal union" (363). But that was all I said of how or why this tactile pleasuring could truly constitute interpersonal union.

Since then, women—fellow scholars, students, and others—have opened my resistant mind further. It is obvious why the tactile pleasuring of making love and sharing meals is a real interpersonal union and thus a profound loving and knowing of the other person. The pleasuring of sex, food, and drink is a "mutuality." We do not have words to name or define it, but when we studied the mutuality that Hadewijch makes central to her experience of God in Part I, we pointed to it with two sets of words. First, the individual acts, is active, and exercises his or her own agency and power. This, I have now introduced above as the "second trait" of the knowing union that the women prize. Second, the individual exercises such power on another individual while the other individual does the same to her. One may or may not want to use here, even loosely, words like "act" and "activity." But the individuals affect each other and meet each other's needs. They correspondingly receive from each other and depend on each other.

Such mutuality I introduced above as one form of the second trait of the devout knowing that medieval women believe they gained in their devotion and treasured. I connect now the knowing in mutuality with the third and fourth traits: It takes place usually out of strong desire. It takes place in pleasuring of common familial kinds. But how does "mutuality" yield its unique knowing by each of the other. The mutual flow of Hadewijch and Divine Love into each other? Their active pleasuring of and being pleasured by each other? The gladly giving and hungrily taking of food and drink? Brother and sister playing together? Spouses, of the same or opposite gender, falling asleep together aided by each other's presence? How does any of this interaffectivity yield a knowing? What kind of knowing? I think hard on it these days, make some stabs (Milhaven, 1991). I do not know. I encourage readers.

In any case, this blind withdrawal of men from pleasurable, familial knowing is a bit of evidence to support the thesis that the traditional Western thinker is a man sitting and pondering alone. If the medieval Pietà epitomizes devout women's experience, Rodin's sculpture *le Penseur* perhaps epitomizes male thought. At my university, a statue on campus is of a man sitting alone on a horse and clad in armor. He is Marcus Aurelius, whose profound writings give little room at the top for interaffectivity with another. I admire this noble intellectual line of Greek descent, still ruling in our own time. I admire it because it is my lineage and because I will defend to the death the greatness of mind of Plato and Augustine and Thomas and Kant and Barth and Rahner, Karl and Hugo. But their reaction in the past and present to the experience we have traced out with five characteristics suggest that, intellectually speaking, 2500 years of male virgin births is enough.

Let me continue the opening of my male thought for impregnation and name briefly two final points of contrast between the experience of the women and the theory of the established theologians. Complete immersion in a singular place and time, agency and interaffectivity, pervasion by physical touch, surging physical need or desire, and pleasuring typical of family life are not the only traits of bodily experience had by devout women which are generally ignored or belittled by the theologians. Pain often is too.

Both physical and personal pain run through much of the rapture and devotion of the women. In the hopeless longing of a mother holding her dead son. In the fear of death, the weariness unto death, of a man who trod the winepress of his suffering alone. Through thorns pressing into the head and a desolate loneliness. One knows oneself uniquely in experiencing that pain. One knows another uniquely in sharing that pain. Imagine, again, the experience of the devout before these late medieval portrayals in the Low Countries and Germany of the passion of Jesus or Mary or the disciples.

The final trait has been seen already in connection with the others. Much of these elemental experiences of urgent human need and passion, climaxing in overwhelming pleasure or pain, is empathic identification, both physical and personal. Bodily identification flows in the women's experience, and is, once more, a mode of interpersonal knowing that the theologians generally ignore or depreciate. The women staying with Christ in his passion and death become one with him through suffering. The women weep with Mary sorrowing. They laugh with the joy of the Baby in their arms. They know.

Once more, too, we recognize a mode of knowing typical of family life. The parent feels the sick feelings of the child. The nursing mother smiles, feeling how the baby experiences her in the feeding. The baby laughs, echoing her generous joy. So too, the sick person and the caregiver. The couple living together after a while take on each other's moods. The person setting the food on the table and the person picking up the food eagerly. They all feel physically the other's feelings as their own. By empathy one knows others in their experience because their experience becomes one's own.

Thomas Aquinas and Aristotle knew well that identification was essential to love. Lovers consider their loved ones as their second self. One considered the other's good or evil as one's own (ST I–II, 28, 2; I–II, 32, 5–6; II–II, 25, 7; I–II, 30, 1–2; Aristotle, Nicomachean Ethics, VIII and IX). But for such traditional thinkers, only the nonbodily identification by reason and will was the love that counted. Bodily, sensory identification was not. In intrinsically good human love, sensory identification, or embodied empathy, was a natural, appropriate, and often praiseworthy complement of the rational identification, but it was not essential to that love. That sensory identification could be inextricably constitutive of intrinsically worthy love, as it was for women of the late Middle Ages, was unintelligible to these men (ST II–II, 30, 3; II–II, 24, 1; II–II, 25, 7 and 12; II–II, 26, 5). Typical is Thomas's emphasis on *rationabile* in endorsing Aristotle's explanation of why parents love children and vice versa (In Decem Libros…, VIII, 12).

Caroline Bynum suggests that medieval women, rather than men, became absorbed in Christ as Eucharistic bread and wine because giving food and drink was a major, ongoing experience of women in that time (1985b, 10–16; 1984, 192–99). I would like to expand the suggestion. The experience of women of those times, even more than today, was centered in family life: not just in putting food and drink on the table, but in giving suck and caring for babies, in playing, in hugging, in helping clothe, in sex, in sympathizing with and caring for the hurt or ill or dying, etc. A certain number of the women whom Ziegler and Bynum cite were or had been married or never left their parents' home. This was particularly true of the Beguines, whose religious communities were everywhere through the Low Countries in the fourteenth and fifteenth centuries. In any case, most of the women whose religious experiences we hear of grew up in family life.

The social fact, I suggest, is not the sole cause. The women, as individuals, must have had and exercised a strong, vital humanness to feel fully and prize their familial living, actual or idealized, had or longed for, and to continue it in their religious life. Their waxing personalness opened them to the revolutionary, nonmetaphorical experiences recounted by women mystics and reflected in the art of their time and place. What we saw in the fifth characteristic is true of all seven characteristics of the women's religious experience: each characteristic is an extension of their ordinary experience in family life.

Something parallel happened with medieval theologians, but, again, they went only part of the way. Perhaps one cause was that for centuries practically none of the theologians had lived long in a family. Certainly, as becoming a monk began to be more often the decision of an adult, not determined for a child by its parents, medieval thinkers, led by Bernard of Clairvaux or Francis of Assisi, multiplied, more than ever before, religious metaphors based on the experience of being in love and making love, on mothering and being mothered. Medieval theologians praised more than ever before, and on its own merits, married and family life, including "fleshly love." But no male theologian, to my knowledge, acknowledged that either this or any other of the distinctive everyday experiences of family was interpersonal knowing.

A good deal of modern feminist epistemology, too, grows out of family life. Into Susan Griffin's *Woman and Nature* (1978) went what the author "learned of the necessities of daily life from the women of my family, the work necessary to keep house together and raise children— all that women know of naming feeling while we live in a culture that misnames and mistakes what we experience" (xi; cf. xvi, 188–91, 197–99, 201–2, 207–8; see also Rich 1976, passim). In *Eros and Power*, Haunani-Kay Trask traces out modern feminists' "vision" of "a qualitatively better mode of being and living," a vision that "is evidence of a critical consciousness which arises out of women's particular form of social practice: their erotic/reproductive roles of biological and emotional mothering of children, men and other women," their emphasis "on physical and emotional gratifications that connect 'life' and 'work'; fluid forms and concepts of intimacy that extend far beyond the merely genital" (1986, x–xi; cf. 86–100). Poems of Robin Morgan convey powerfully the precious knowledge she gained from her pregnancy, nursing her child, and caring for her elderly mother, "cleansing her genitals and helping her to urinate." Trask observes that Morgan's poem, "Network of the Imaginary Mother," is a deliberate "stunning reversal" of the ritual and meaning of the Eucharist (138–40). Medieval and modern women cross forbidden lines at the same point and make history in that they, among other things, extend publicly this kind of knowing outside of family life. The similarity of the medieval and modern accounts is all the more suggestive when modern women advocate this kind of knowing not only for religious life but also for secular sisterhood which includes lesbian relationships. Committed lesbian and gay relationships are as "familial" as similar heterosexual couples.

Why, therefore, could the medieval theologians not recognize the widespread religious experience of women as a genuine knowing of Christ, Mary, and so on? Because the experience was too physical to be knowing of a person. It was too physical in at least seven respects: (1) in such knowing the woman knew only an individual person in a single moment and place; (2) she knew a person in actively affecting another, often as the other person affected her; (3) physical touching pervaded the knowing; (4) the knowing arose in bodily need or desire; (5) the knowing was had often in pleasures peculiar to family life; (6) the knowing was had often in bodily pain; and (7) the knowing included often bodily identification with another.

The medieval theologians' opposition, therefore, helps to place in relief seven facets of the bodily knowing that the women reported. The seven make a unity. They do not contradict each other but can be combined to make a single concept of "bodily knowing." The facets overlap but are distinct aspects of the whole. This heptagonal concept is not only logically coherent but points convergently to recognizable experience. Moreover, they are found together manifestly, though not exclusively, in the same milieu of family life.

I conclude Part II with two suggestions for reflection and discussion. First, despite enormous differences between medieval and modern life and thought, the "bodily knowing" experienced, reported, and extolled by certain medieval women and by certain modern women is analogous. The heptagonal idea I have drawn from the medieval accounts seems to me, as I have indicated ocasionally in passing, to outline also the bodily knowing which certain women writers advance today, yet for which they have not, to my knowledge, offered a systematic conceptual overview. The same is true here that, as a conclusion of Part I, I suggested concerning the kind of mutual loving delineated there.

My second suggestion is that the women are right. Not only by reason, but also in and through their bodies do human beings know other human beings in their humanness, their personalness. Not only by reason but also in and through their bodies do human beings know much that is intrinsically precious in human life. The recognition of this fact does away with any medieval or modern epistemology that propounds a hierarchy of knowledge. It demands instead a bipolarity. Some of my discussion partners, such as Emily Stevens, refuse here to speak of bipolarity, though possibly of multipolarity. They are probably right, but bipolarity is where my argument presently stands as I finish off my book by proposing a hypothesis for further reflection.

If human knowing be taken as bipolar, the various forms of human knowledge of person and value do not constitute degrees of realizing one supreme kind of knowing, as the theologians thought they did. The various forms of human knowledge of person and value fan out between two polar kinds of knowing. One is bodily; the other is rational. All human knowing is constituted by its particular degree of participation in both poles. The poles are irreducible to each other. They are incomparable to each other in value. One cannot be rated superior to the other. Good human living is at any moment determined by both poles. Again, this parallels what I suggested as conclusion of Part I.

Since Western culture has confined its light to rational knowing or blind faith, it will make history if Western thinkers locate and investigate where in the shadow the other kind of knowing, or better: the other side of knowing, goes on or can go on. This, I agree with many modern women, will affect areas of human decision and action such as war and peace, social justice, sexuality and reproduction, education, and interpersonal relations generally. It is likely to affect all areas of human decision and action.

I know the fears of male thinkers, trained in dominant Western thought. God knows what we may find if we acknowledge in principle this bodily knowing, ponder it in experience, and exercise it consistently in practice by and beyond hearth and home. God knows what may happen in our lives if we give ourselves as wholeheartedly as we can to this mutual loving. But the alternative, the narrow track chosen by our forefathers, is confining.

# Appendix:
# Thomas Aquinas on the Pleasure of Sex and the Pleasure of Touch

In this essay, which I wrote much earlier than the body of the present book, I trace why a classic male theologian of the Middle Ages denied any intrinsic value to spouses' bodily enjoyment of sex. I outlined most of this tracing in Chapter 10, but the further argument and detail in this appended essay should bring out more clearly and extensively the operative presuppositions of Thomas. It should thus set further in contrasting relief the presuppositions of the medieval women, listened to in this book, about the value of embodied mutuality, particularly about the value of knowing through touch and out of bodily desire. On the other hand, the limits of the lines of my tracing fifteen years ago, in contrast with what I do now in the body of this book, may, I hope, illustrate the difficulty for a rationally trained mind, "even" a modern one, to keep at digging up the presuppositions, his as well as the author studied. For this reason, I reprint today this essay with only minor editing.

# 1

The moral appraisal Thomas makes of sexual pleasure is central to his whole sexual ethics. Like the rest of his special ethics, his sexual ethics is framed in terms of virtues and vices.[1] The basic virtues and vices pertinent to sexual ethic—*temperantia, intemperantia, castitas, luxuria, virginitas*—are defined in terms of their relation to sexual pleasure. Thomas does not define them, as one might expect him to do, in terms of their relation to the right kind of sexual activity or to the proper end of sexual activity. He speaks often enough of right and wrong kinds of sexual activity and of the proper end to be sought in sexual activity. But the basic virtues and vices pertaining to human sexuality consist formally in the right and wrong dispositions toward sexual pleasure.[2]

Even sympathetic modern commentators (Fuchs 1944, 24–26, 55–56, 60, 66–71, 219, 226–27, 273; Doherty 1966, 278–79; Noonan 1965, 293–95; Van der Marck 1967, 105) have judged that the various statements Thomas makes about sexual pleasure do not form a coherent, consistent whole. In certain passages, Thomas evinces a negative evaluation of sexual pleasure; in other passages, a positive one. I will, however, try to show that Thomas's total view of sexual pleasure is a coherent and consistent one. On the other hand, I will eventually suggest that his view is a narrow, limited one that invites the critical, constructive dialogue of the contemporary Christian ethicist.[3]

The present essay centers on the only sexual pleasure that Thomas did not consider to be necessarily sinful, i.e., the sexual pleasure of married people. At first glance, Thomas appears to be inconsistent in his comments on this pleasure. On the one hand, when he lays down certain moral rules for married Christians, he seems to have a low opinion of their sexual pleasure. Adducing the nature of sexual pleasure, he prohibits or discourages sexual intercourse at certain times in the lives of the Christian spouses. Their sexual relations render inappropriate the reception of the Eucharist on the following day

(ST III, 80, 7; ST III, 64, 1). They are not permitted on holy days when one should devote oneself to prayer and spiritual matters (64, 5). They are not appropriate during times set aside for religious meditation or religious services or liturgy (41, 3).

These strictures coincide in part and differ in part in what is discouraged or forbidden. But Thomas's reason for each stricture is the same: the nature of sexual pleasure. In humanity's fallen state resulting from original sin, sexual pleasure is not submissive to reason but, absorbing the mind, distracts it from spiritual realities. Although conjugal sex, generally speaking, is not sinful, the consequences of its pleasure make it unfitting for times given to spiritual activity.[4]

In another passage, Thomas darkens further the picture of sexual pleasure. He is advancing an argument why close relatives should not marry.

> The pleasure of sexual intercourse "seriously corrupts the judgment of moral wisdom." To multiply such pleasure, therefore, is contrary to good morals. Now such pleasure is increased by the love of the persons in the union. It would, therefore, be contrary to good morals for close relatives to be united, because in their case the love that comes from common origin and sustenance would be added to the love of concupiscence. And, the love being multiplied, the soul would be necessarily more overcome by the pleasure.[5]

Thomas's view of natural human love, expressed in the passage just cited, is curious. Why does he not discuss the possibility that "the love that comes from common origin and sustenance," i.e., a love natural among family members, might also make some positive contribution to the higher activities of humans, including their moral wisdom?[6] What is to our present point, however, is that Thomas sees conjugal sexual pleasure as corrupting moral wisdom. Here, as in the strictures previously considered, Thomas premises that sexual pleasure has considerable disvalue because of its effects on the mind of the spouse.

Other passages convey a still more negative view of conjugal sexual pleasure. In these texts, the premise for Thomas's argumentation is not the deleterious effects of sexual pleasure, for he pronounces on all and any conjugal intercourse. Some of this intercourse is necessary for procreation and this presumably justifies any bad consequences. But whenever it is the appetite for "sexual pleasure" (*concupiscentia*) that moves the husband and wife to have intercourse with the other, they sin, at least venially. They sin, therefore, not because of what they do or because of the results of what they do, but because of what moves them to do it. They sin because they are acting "out of" (*ex*) their appetite or desire for sexual pleasure.[7]

This condemnation of the desire for pleasure as motive for conjugal intercourse has drawn criticism from modern readers of Thomas. Moreover, it seems to contradict the positive utterances of Thomas that we are about to consider. However, the burden of our argument will be that this condemnation is no inconsistency but if rightly understood, reveals the coherent rationale of Thomas's systematic treatment of sexual pleasure and, therefore, of Thomas's whole sexual ethics.

---

## 2

---

In apparent conflict with the negative statements just considered, Thomas's moral appraisal of fallen humanity's appetite for sexual pleasure is a relatively positive one in comparison with that of other medieval theologians and later theologians of the Reformation. He rejects the widespread medieval position that this "appetite" (*concupiscentia*) is essentially sinful. He also denies, and is the first medieval theologian to do so, that this appetite constitutes a flaw or perversion of human nature. Original sin has "corrupted" the appetite only in reducing it to its natural state. In its natural state, *concupiscentia* is no longer completely subject to reason. Instead, it tends simply toward its own object, sexual pleasure, and can be governed by reason only to a limited extent. But this spontaneous tendency is the "state befitting man according to his natural principles." The total subjection of a person's lower forces to his reason, the state he enjoyed before the fall, was due not to his nature, but to the "original justice given him over and above his nature by divine liberality." Christian virtue need not and cannot regain this "supernatural" dominion over man's lower nature. The virtue of chastity can coexist with this "penalty" (*poena*) of original sin, i.e., with the purely natural reality of its sex appetite.[8]

Indeed, the virtue of chastity requires this natural sense appetite. Deficiency in this particular sense appetite, i.e., a certain insensibility to sexual pleasure, constitutes one of the vices opposed to the mean of chastity.[9] Chastity is a virtue not merely because of restraint but also because of promotion of sexual activity. Chastity entails, for the married person, the sexual activity that fulfills the goal of marriage: procreation.

This activity properly proceeds from two appetites: the sexual appetite, which is one of the sense appetites, and the general rational appetite, which is will. Consequently, the virtue of chastity disposes and orders both of these appetites to the appropriate activity. If the sense appetite, i.e., the individual's sexual passions, is not habitually disposed and ordered in this way, he is not chaste. He or she is not

chaste—even if the rational appetite is well developed enough to with-
stand the passions and govern sexual behavior so that he or she acts in
an exemplary way.[10] A morally good bodily act calls for the contribution of a sense appetite as well as of the rational appetite. "Just as it is
better that a man both will the good and do it in an external act, so, too,
it belongs to the perfection of a moral good that the man be moved to
thc good not only by his will, but also by his sense appetite."[11] The
virtue of chastity requires, therefore, that the appetite tor sexual pleasure move the chaste spouse to intercourse.

Has Thomas contradicted himself? *Concupiscentia* is not sinful. It
is not a flaw or perversion of human nature. It is fully natural and
should move the chaste spouse to conjugal intercourse. On the other
hand, as we saw earlier, the spouse sins if *concupiscentia* moves him or
her to conjugal intercourse. The contradiction, however, is only apparent. As Joseph Pieper (1965, 153–75) points out, the impulse given by
*concupiscentia* is understood differently in the two statements.

The impulse of *concupiscentia* required by the virtue of chastity is
one approved by a person's reason. The impulse that is sinful is one
lacking any such approval. The spouse sins if his or her sexual appetite
on its own, without the concurrence of reason, moves him or her to
intercourse. The spouse acts virtuously if the appetite does this under
reason's governance. In condemning pleasure as a motive for conjugal
intercourse, therefore, Thomas does not contradict his assertion of the
moral necessity and value of the sensual desire for sexual pleasure. He
is only insisting on the even greater moral necessity and value that reason, seeing reality lucidly, be in command.

In this context, however, Thomas insists also on something else,
which Pieper ignores. In approving the exercise of the sexual appetite,
reason cannot have the motive of sexual appetite. It is essential to
virtue not merely that reason approve the exercise of the spouse's sexual appetite. It is equally essential, says Thomas, that that exercise
never be "for the sake of pleasure alone" (*propter solam delectationem*).
Reason can never approve such a purpose. The sexual appetite, itself,
being a sense appetite, necessarily seeks its pleasure, as it moves the
spouse to engage in intercourse. But pleasure is not the "end" (*finis*)
ordained by nature for sexual activity. Consequently, the virtuous sexual activity of the spouse must be ordered by reason to some other
end.[12]

Thomas clarifies the principle in a further application of it. It is
contrary to virtue to refrain from intercourse now in order to enjoy it
more later. The rational agent would be "intending" the pleasure as at
least part of one's "end in view" (*finem*). He or she would be seeking the
pleasure "for its own sake" (*propter seipsum*). Reason cannot do this.[13]

We see here why Thomas condemns marital intercourse had "out of sexual desire." His reason is not any intrinsic evil in the desire or harmful consequences of the intercourse. Nor is Thomas simply condemning those instances when, because of circumstances, reason does not endorse the satisfaction of the desire. In condemning marital intercourse "out of sexual desire," Thomas condemns any instance when the spouses have sexual pleasure as the purpose of their intercourse. He does so because that is a purpose reason can never approve of.

This is confirmed by the larger context of the chapter in *Summa Contra Gentiles* we have been examining. The central question of the chapter concerns the afterlife of the blest once they have been reunited with their bodies. Will the blest then engage in sexual activity? In that beatitude, Thomas tells us, God will make the body totally subject to the rational soul. All bodily appetites, actions, passions, movements, and so on of the individual will be rendered perfectly submissive to the soul. They will, therefore, in no way be able to impede the higher life of the blest. They can have no harmful consequences in a person's spiritual life.[14] And yet the argument that the blest should enjoy sexual pleasure "lest any pleasure be lacking in their ultimate reward" is rejected by Thomas.

Even though it would have no harmful effects on a person's spiritual activity, sexual pleasure may not be sought for its own sake. Thomas's negative judgment on sexual pleasure is, therefore, not based exclusively on the power this pleasure has, as a result of original sin, to move people out of the control of their reason and thus to disturb their higher life. A more fundamental negativity of sexual pleasure is seen by Thomas even when the pleasure is totally subordinated to a person's reason and higher life and his or her intellectual contemplation goes on undisturbed. This is neglected by even excellent commentators such as Bailey (1959, 134–38, 243). It is the principal subject of the present essay.

Does Thomas, therefore, see something morally evil in sexual pleasure? He has affirmed that the sense appetite for sexual pleasure is perfectly natural, sinless, and essential to virtue and to the perfection of moral good. But might he not hold, at the same time, that sexual pleasure itself is intrinsically evil? This would explain why it could not be the intended purpose of a person acting rationally.

In opposition to earlier theologians, however, Thomas denies that sexual pleasure is a moral evil. He implies that it is morally good. This is reflected in Thomas's teaching that virtue need not restrain the amount of sexual pleasure which the spouse permits himself in intercourse. Virtue lies in keeping the mean; the mean chastity observes lies between the vices of "license" (*luxuria*) and insensibility. The chaste spouse, therefore, *should* enjoy the pleasure of intercourse.[15] Moreover, the virtue of chastity does not restrict the amount of sexual pleasure permitted the spouse in intercourse.[16] The mean of virtue is not a quantitative one determining an amount of sexual pleasure that is neither too much nor too little. No "abundance" or "superabundance" or "extreme intensity" of pleasure in conjugal intercourse offends chastity or makes the intercourse sinful.[17]

In another context, too, Thomas implies that sexual pleasure in itself is not just not evil but positively good.[18] He affirms that, had there been no Fall, "This sense pleasure would be so much the greater as human nature would be finer [*purior*] and the body more sensitive [*sensibile*]" (ST I, 98, 2; cf. Fuchs 1949, 26–27). Moreover, God endowed sexual activity with pleasure in order to motivate humans to this activity essential for the continuation of the human species.[19] If virtue obliges the spouse to enjoy his or her sexual pleasure and if a superior human nature would have even greater sexual pleasure than a person can have now and if God has made the pleasure such as it is so that people may desire it, it must be good in some real sense of the word. How, then, is it wrong for an individual acting rationally to seek it for its own sake? Why cannot human reason approve conjugal intercourse for the purpose of this pleasure?[20]

# 3

The statements of Thomas that we have been examining fit together when located in the general framework of his complex but coherent anthropology. The key is to distinguish all along the line between a person's sense appetite based on sensation and his rational judgment based on reality. Thomas makes his more positive statements about sexual pleasure when he is considering the sense appetite by itself. He makes his more negative statements when he is concerned about the judgment of reason and the full reality on which it must be based.

By its nature, the sense appetite seeks only one good: sense pleasure. This pleasure is a real good, and it is good and necessary that the appetite seek it. This is God's and nature's intent. The sex appetite, therefore, of the virtuous spouse should seek its pleasure for the sake of its pleasure. In this sense, one can say that the sexual appetite properly seeks pleasure as its "end" (*finis*) and moves the person to act "for the sake of sexual pleasure" (*propter delectationem*).[21]

Nevertheless, though the sense sex appetite seeks sexual pleasure as its end, that pleasure is not its end. Sense appetites simply follow sense perception, and sense perception does not attain reality or what is. It is therefore quite consistent to hold that the real end of the sex appetite is different from the end it seeks, pleasure. The real end of the sex appetite is what God and nature have in view in affixing pleasure to the appetite as an inducement for the appetite to move. The end God and nature have in view is the conservation of the human species through procreation. For nature and God and, therefore, in reality, the pleasure that this sense appetite seeks is not its end, but only an "instrument" to lead it to attain its true end, procreation.

The conservation of the species, in turn, has as its ultimate end the ultimate end of the members of the species, i.e., the spiritual activity by which they have their beatitude with God. The senses can know nothing of ends like these. Conservation of the species is not perceptible or imaginable by the senses. Neither is any spiritual activity of humanity. Consequently, no sense appetite can seek these goals. The sex appetite thus contributes to its true end unknowingly, as it seeks its pleasure as its end. But a person's rational appetite may seek knowingly only the true end of sexual intercourse, realized in procreation.[22]

*4*

At issue, therefore, is not Thomas's consistency but his dualism. All anthropologies are dualistic or pluralistic; no one finds the human being to be a perfectly simple being of one piece. But Thomas's treatment of sexual pleasure, as seen thus far, seems to suggest an underlying dualism whereby a person's sense life would have no value except in serving his or her spiritual life. This is difficult to defend in terms of either ordinary human experience or traditional Christian dogma.

Is not the body with its sense appetites more than a mere instrument of the rational soul? Does not reason know that the rich sense life of an individual is worth seeking for itself and not solely to make possible his or her spiritual activities? Does not the Christian belief in the resurrection of the body require that the real ends of human activity and the proper purposes of rational humans include the sensual bodily as well as the spiritual? In the light of Thomas's teaching on sexual pleasure seen thus far, one would expect Thomas to reply negatively to these questions.

Instead, Thomas gives an unequivocally affirmative answer. All the senses that humans have they have in common with brute animals. One purpose of a person's senses is the same as that of the brutes': to help acquire the necessities of life for the individual and the species, mainly, food, drink, and sex. But human senses have another purpose, not shared by any brute animal. The senses have been given also for the knowledge they yield. The knowledge to which the human senses are ordered is primarily the rational knowledge they make possible. All rational knowledge must take its beginning from sense knowledge. But the senses are also given to humans for the sake of sense knowledge itself (ST I, 91, 3).

Sense knowledge is, in itself, "good" (*quoddam bonum*) for a person. A rational person knows this and loves his or her senses "for" (*propter*) the sense knowledge they give. Unlike the brutes, and by virtue of reason, a person takes pleasure in his or her sense knowledge, i.e., both in the sensible objects known and in the act itself of

135

knowing.[23] Sense knowledge is, therefore, itself, an end of humanity. Humans rationally endorse this end and acts for the sake of it. In Thomas's ethical anthropology, it follows that the same is true of the pleasure people take in their sense knowledge. That pleasure, too, is, itself, an end of humanity, and people rationally act for the sake of it.[24]

This evaluation of sense knowledge in general and people's pleasure in it is diametrically opposed to Thomas's evaluation of a person's sexual experience in particular and its pleasure. The same opposition is reflected also in Thomas's conception of the afterlife. Although, as we saw, there will be no sexual pleasure in the afterlife, still all the bodily senses of the blest will be active and give them knowledge and pleasure (Suppl., 82, 3 and 4; 83, 6; ST I–II, 3, 3). The beatitude of the blest does not consist formally in any bodily good and, therefore, not in this sense knowledge and pleasure. Human beatitude is essentially the immediate intellectual vision of the uncreated good. Essential human beatitude is, therefore, purely spiritual. Moreover, in the beatific state, an individual no longer needs sense knowledge to make possible any of his or her intellectual knowledge, as he did in the terrestial life (ST I–II, 2, 5; 3, 3 and 8). Nevertheless, in the state of beatitude human nature will be in the "greatest perfection" (*in maxima perfectione*), and this perfection necessarily includes the appropriate activity and pleasure of the senses (SCG IV, 86; Suppl., 82, 4; 83, 6).

Thomas puts the same reasoning in yet another form that brings out sharply the opposition between his appraisal of sense activity in general and his appraisal of the sense activity of sex. Until the bodies rise from the earth at the last judgment and join their souls in beatitude, the "natural appetite of the soul for its full perfection and happiness" (*naturale hominis desiderium ad felicitatem*) is not completely at rest. Since the soul is by its nature form of the body, its basic appetite "desires" (*appetit*) and "wants" (*vellet*) that the body share its "enjoyment" (*fruitio*) of God (SCG, IV, 79; ST I–II, 4, CT 1, 151). The rational soul, therefore, has a general appetite for sense activity and pleasure for their own sake. And yet as we saw, the rational soul has no appetite for the sense activity and pleasure of sex for their own sake.

Another argument of Thomas illustrates the same contrast. He argues that the blest will have full sense activity and pleasure because their body and senses must also have their eternal reward (Suppl. 82, 4). This is exactly the same reasoning which, as we saw, Thomas rejected when applied to the sense activity and pleasure of sex.

---

In sum, Thomas takes two contrasting positions on sense pleasure. In general, an individual rationally acts for the sake of sense pleasure; it is an end for his or her rational appetite. A person does not act rationally for the sake of the sense pleasure of sex; it is not an end for his or her rational appetite. Thomas clearly considers both positions essential to his ethics. He maintains each on a good number of occasions and in different contexts. Thomas is aware of the contrast between the two positions; in fact, he underlines it. The pleasure of sex is essentially different in its moral goodness from most of the pleasures that the senses yield. The question is: What is the difference?

To understand the difference, it is worth noting that sex is not the only sense activity different in this regard from sense activity in general. Eating and drinking receive the same negative judgments as sex. It is wrong to eat or drink for the pleasure of it; the pleasure of eating and drinking is not an end for human activity.[25] Consequently, it is contrary to virtue to abstain from eating and drinking at a given time in order to enjoy it more later (SCG IV, 83). In a person's beatific life with God after death, he or she will not engage in eating or drinking or have their pleasures (SCG III, 27; IV, 83; Suppl. 81, 4; ST I, 97, 3).

That the pleasures of food and drink receive from Thomas the same moral treatment as the pleasures of sex is no accident. They have certain characteristics in common, distinguishing them from the other sense pleasures. The pleasures of sex, food and drink are peculiar pleasures of thc sense of touch.[26] Their purpose is directly to serve nature, i.e., to incite the appetites of the animal (whether brute or human) to seek activities which are needed to preserve the individual and the species. As a result, these pleasures have been made so powerfully attractive that humans need a special virtue to govern them. The sole function of the cardinal virtue of "moderation" (*temperantia*) is to govern the desires and pleasures of touch connected with sex, food, and drink.[27] Finally, of the "goods" (*bona*) sought by a person's different sense appetites and passions, those giving the pleasure of touch basic to sex, food, and drink are the "lowest" (*infima*) (ST I–II, 60, 5; cf. NE III, 10, 17ᵇ 23–18ᵇ 8).

This last assertion by Thomas suggests an avenue for exploring further our present question: What is the difference, in intrinsic moral goodness, between the sense pleasure of sex (to which we now add food and drink) and the general pleasure that the senses give people? Why does Thomas rate the latter essentially superior to the former? In principle, the intrinsic moral goodness of a pleasure is determined by the "good" (*bonum*) which gives the pleasure, i.e., by the particular good in which pleasure is taken.[28] Since the pleasures of sex, food, and drink have been judged inferior to the generality of sense pleasures, then one is not surprised to hear that the goods in which the pleasures of sex, food and drink are taken are of the "lowest" (*infima*) sort. But what makes them the lowest?

In the other sense pleasures, as we saw, sense knowledge is the good in which people take pleasure. The different kinds of sense knowledge are, in turn, rated as more or less good according to the degree of knowledge they give.[29] An internal sense is superior to an external sense because its object "is known more."[30] The external sense which a person values most for itself is the sense of sight, for this external sense is "more knowing."[31] The senses whose object people call "beautiful" are those which are "the most knowing," i.e., sight and hearing.[32] If, therefore, the goods in which the pleasure of sex, food, and drink is taken are of the lowest sort, any sense knowledge contained therein must be of a minimal degree of knowledge.

Thomas has told us that the pleasures of sex, food, and drink come from the knowledge by the sense of touch that the appetite now possesses its object. But no text of Thomas to my knowledge affirms that this particular sense knowledge is to any degree good in itself or that a person takes any pleasure in it as knowledge. One understands then why he rates the pleasures of sex, food, and drink as, in themselves, essentially inferior to all other sense pleasures. The sense of touch when it gives rise to the pleasures of sex, food, and drink must be, to use Thomas's quantitative terminology, so little knowing that the knowledge it gives has no intrinsic goodness worth the consideration of rational people.[33]

The crux of the matter, therefore, in understanding Thomas's appraisal of sexual pleasure, is his failure to accord more than a minimal, negligible kind of knowledge to sexual experience. Even in his historical context, this epistemological lacuna is curious. It runs counter to familiar biblical language where "knowing" is a synonym for sexual union (Gen. 24:16; Luke 1:34, cf. Thielicke 1964, 66–67). The corollary of this incomplete epistemology—namely, that sexual union is, of itself, a blind, purely subjective experience and therefore considered simply in itself of minimal human value—runs counter to the familiar biblical understanding of sexual union as profoundly interpersonal (Gen. 2:24; Matt. 19:3–6; Mark 10:2–9; 1 Cor. 6:13–20; Eph. 5:31–33). The resultant view of conjugal sex runs counter to language frequently employed by medieval mystics, such as Bernard of Clairvaux and Mechtilde of Magdeburg (see O'Brien 1964, 103–4, 119–20) in which the sexual union, knowledge, and pleasures of spouses are seen as a worthy analogue of the most intensely, profoundly, and sublimely personal union, knowledge, and pleasure possible to humans on earth. A similar high esteem of conjugal sex is implied in the allegorical interpretations of the Song of Songs, "the book most read and frequently commented on in the medieval cloister" (Leclercq 1961, 106; cf. e.g., Song of Songs 4–5:5; 7:1–13).

Historians of Christianity and Christian dogma, psychosocial historians of the Middle Ages, cultural anthropologists, and others may be able to probe more deeply the causes of this blind spot of Thomas's epistemology. But the historian of Western religious thought can also elucidate, for Thomas, as we have seen, expounds his sexual ethics philosophically. Thomas's low opinion of any knowledge given by sexual experience is of a piece with his whole philosophical scale of human values. The sole norm, ideal, model and crown of all human values is human reason (cf. NE I, 7–8; X, 6–8). Sexual experience, of all human experiences, is the least like human reason.

One has noted the quantitative terminology Thomas employs in evaluating kinds of sense knowledge: "is known more," "more knowing," "the most knowing." To be able to express relative value simply in terms of "more" or "less" presupposes a single scale of values, determined by a single norm of fullness or perfection. For Thomas, that norm is "reason." A given kind of sense knowledge will be more or less knowing to the degree to which it "shares in," i.e., takes on some of the attributes of the knowledge of reason.[34]

Like any two neighboring levels of the hierarchy of being, sense knowledge and rational knowledge cannot be contrasted in simple, blanket fashion. In its higher reaches, a level of being approximates more and more to the next highest level in nature and worth. Subspecies of the lower level have more intrinsic value to the degree to which they are more similar to the higher level. Of the different kinds of sense knowledge, that involved in the pleasures of sex, food, and drink shares and resembles least the higher rational level. It is the "furthest from reason."[35] In these pleasures, "is seen least of all the brightness and beauty of reason" (ST II–II, 142, 4; cf. 141, 2 ad 3; NE X, 5, 75[b] 35–76[a] 3). Indeed, the whole process of sex appetite, knowledge, and pleasure, unlike other sensory processes, has no share in reason at all, but is totally sense.[36] This is the reason why, of all sense pleasures, the pleasures of sex, food, and drink most immerse a person in the sensible and draw him or her away from intelligible reality.[37] With most of these statements of fact, i.e., of the practically complete arationality of the pleasures of sex, food, and drink and the knowledge involved therein, the context makes clear that they are *eo ipso* judgments of value, i.e., of the inferiority of this kind of pleasure and knowledge.

The cause why sexual pleasure is the lowest of human pleasures and lacks all intrinsic value for the person is thus clear. The cause is not that sexual pleasure interferes most with the exercise of human reason. It does so interfere in a person's purely natural state, but the cause lies deeper. Indeed, we saw that in the afterlife, as in Thomas's hypothetical paradise without the Fall, God could eliminate the interference and sexual pleasure would still lack all intrinsic value. The crucial reason for sexual pleasure's lack of intrinsic value is that it has in it nothing resembling rational knowledge. Unlike higher sense pleasures, it is grounded in pure sense knowledge that has no share in reason.

This judgment of fact is all the more crushing a judgment of value because the value of reason, in turn, lies in its being a participation of the knowledge had by the higher, purely spiritual beings and ultimately had by God (DV 15, 1; In II Sent., 16, 1, 3; In II Sent., 35, 1, 4; In X Eth. Nic. 11, 2110; VI, 4, 1807. Cf. NE X, 7 and 8). God's own knowledge, therefore, is the ultimately determining norm, ideal, and model of all human knowledge and thus of all human values. To say that the knowledge had in sexual experience is most unlike the rational is to say that it is most unlike the divine, the participation in which gives all meaning and worth to human life.

# 6

The crux of the matter, therefore, in understanding Thomas's appraisal of sexual pleasure is his exaltation of the rational as the sole norm and model for all human values. But what does Thomas mean by the "rational"? Luther, Galileo, Descartes, Newton, Rousseau, Kant, Hegel, and many others have passed this way since Thomas's time. The modern understanding of "reason" is not likely to be identical with Thomas.

There is no place left in the present essay to undertake to trace what Thomas means by his all-holy reason. Let me, however, point to one set of clues appearing in the texts dealing with the value of sexual pleasure. Thomas frequently asserts the inferiority of the pleasures of sex, food, and drink on the basis of the fact that they, unlike other human sense pleasures, are enjoyed by humans in common with brute animals. This fact follows, of course, from the arationality of these pleasures, since humans are "the rational animal." But at times Thomas does not mention the arationality of these pleasures. He adduces simply the fact that they are had in common with brute animals and this suffices to prove their inferiority and lack of intrinsic value.[38] Correspondingly, the sense pleasures worth having for their own sake, e.g., in the afterlife of the blest, are pleasures, which, Thomas makes clear, animals do not have (SCG III, 27 and 63). Animals do not enjoy beautiful sights or sounds or smells or tastes. The content of the all-determining norm, reason, is derived negatively from observation of what brute animals do as well as positively from observation of what people do and brutes cannot.

Parallel to and intertwined with the brute animal, a second negative face of Thomas's supreme norm, reason, is easily seen: the human being unformed by civilization. In his value judgments, Thomas often substitutes for, or adds to, the criterion "man as man" (or "reason") versus "the animal." The criterion he adds or substitutes is "man as man" (or "reason") versus "the child," "the boy,"[39] "the fool,"[40] "the slave,"[41] "the peasant,"[42] or even "most humans."[43] The way in which

141

the unformed, uncivilized masses think, feel, and determine their actions is little different from the life of the brutes. It does not compare in worth to the life of that relatively rare human being, the fully developed and civilized adult.

One understands why Thomas treats as essentially superior to the pleasures of sex, food, and drink the pleasures of seeing beautiful forms and beautiful colors, hearing sweet melodies, smelling excellent fragrances, savoring sweet or combined tastes, and experiencing the bodily pleasure had in gymnasium sports.[44] They are the pleasures of the civilized man. Brutes, children, the uncultured, and the masses are not capable of such pleasures. Conversely, pleasures they are capable of cannot be really worth pursuing. One understands why the spouses having intercourse for the sake of its pleasures sin at least venially. They are letting themselves act on the kind of motive that moves brute animals, children, and the unformed masses. To do these things the others do is no evil. To do it for the motive these others have is to fall indeed. It is to fall below the truly human level.

When, therefore, Thomas prizes a particular kind of human knowledge or a particular kind of human pleasure because it has a certain share in reason, we know something of what he is thinking. Behind the word "reason" in the forefront of his mind stands the ideal adult envisaged by the civilization and education of his time. It stands out in contrast to the childish barbarism and animality from which medieval civilization had emerged, which still infested that civilization, and against which the civilization continually struggled. Yes, for Thomas as for Aristotle, the ideal stands out in contrast to "ordinary people" (*plurimi*).

# 7

What we have just touched on is, of course, far from exhausting what Thomas means by reason and what he finds so valuable there. We are still just starting to explore the Thomistic concept of reason when we add other well-known traits of Thomistic reason: clarity; distinctness; recognition of order and proportion; grasp of the necessary, universal, and eternal as opposed to the contingent, singular, and temporal; and so on.[45] Nevertheless, all these characteristics suggest a hypothesis with which I would like to conclude. I take it that a historian of Western religious thought may do more than record past thoughts. He may choose to enter into dialogue with a past thinker. Max Muller (1958, 141–42) has remarked:

> Dialogue with a thinker thinks about what he shows and what he hides, what he has to hide precisely in carrying out his task of showing….one has to show…what had to remain hidden to such a thought. It may well…have justifiably remained hidden at that time because of…what was coming to light. But today it may he precisely what has to be thought about, perhaps, even *the* task and *the* salvation.

What does Thomas have to hide precisely in carrying out his task of showing? Might one thing be a second model of knowledge and thus a second, autonomous norm of values? Could one argue the following? Thomas's model of knowledge is useful, true, and incomplete. Reason, in Thomas's sense of the word, is a valid model, ideal and norm of human knowledge, but not the sole one. A given kind of human knowledge, therefore, is not to be evaluated exclusively according to the degree to which it approximates this single model. Human beings also have a second kind of knowledge that has its own value and, in its fullness, is a model, ideal, and norm in its own right. The two kinds of knowledge are essentially different, each valuable in its own way, not comparable, but complementary to the other. For example, a human being can know in two essentially different ways mothering love: the imminence of punishment and his or her personal autonomy. The two ways of knowing these human realities each have their own value, not comparable but complementary to the other.

The second kind of knowledge, the kind Thomas did not show, is had by the civilized adult, but it is not knowledge that civilization has taught him or her. It is like the knowledge had by children, the uncivilized, the "masses," and even animals, so far as we can judge. It is unlike what one calls "rational" knowledge. What is known in this second way is known dimly and obscurely, narrowly singular, immersed in time and place. Although it does not have all the limitations Thomas attributes to sense knowledge, it *is* sense knowledge or at least is indistinguishably interfused with sense knowledge. The sense most often operative is the sense of touch, paradigmatically exercised in sex, food, and drink. The second kind of knowledge resists, though not with complete success, the first kind's (reason's) efforts at clarifying and analyzing it. The hard-won rational schemata, concepts, and words can point to and to some extent bring further into the light the original, elementary experience. They can distinguish some of its elements, but they cannot take their place.

What Thomas does not and in his time could not show is not the fact of this kind of knowledge, but its autonomous value in human life. Recognizing only some of its characteristics, he places it at the bottom of his scale of values. For him it constitutes a purely negative pole, furthest from the single model and norm of all values and good. He fails thereby to recognize that this pole is not a negative one but a second positive one. People have two, contrary kinds of knowledge, each with its own internal value. Compare two particular acts of knowing, e.g., a physicist's knowing physics and knowing her spouse's feelings toward her at a given moment. One cannot simply say, as Thomas would, in knowing this, the physicist knows more than in knowing that. At most, one might be able to say that in knowing this, she knows more with one kind of knowledge; in knowing that, she knows more with the other kind. Even this, though, could be misleading since almost all human knowing, and especially the best, is not centered at either pole, but in a tension and interrelationship, a balance between the two.

A loving, sexual union yields eminently this second kind of knowledge. The pleasure of sexual love is worth seeking because, among other reasons, it involves this kind of knowledge.[46] Today one might prefer to call it not "knowledge" but "experience," "awareness," "openness," or "communion." What is essential is that one does not understand and assess it using as the sole model some kind of knowledge epitomized by one's civilization.

I suggest all this as a hypothesis for critical discussion. Whether it has any truth to it can be determined only by the critical discussion. But true or not, it may give some focus to current discussion of sexual pleasure in Christian ethics. I have argued elsewhere that that discussion is in disarray, with unattended confusions and strange silences. In invoking the value of nonrational experience, contemporary Christian thinkers may well be showing something Thomas could not. But they are not showing it as well as he showed what he did show (cf. Milhaven 1976 and 1974). Current ethical discussion of sexual pleasure would gain if dialogue with Thomas inspired them to emulate his lucidly probing, rigorously coherent synthesis even as they widened its perspective.

# NOTES

## Introduction

1. My students have lustily and shrewdly compelled me to ask more clearly and listen harder than I would have otherwise. I am grateful above all to Collette Ah-Tye, Diana Cates, Kimberly Colwell, Christine Jonas, Robert Newman, Carmen Parcelli, Jennifer Seltz, and Emily Stevens.

## Chapter 1

1. See the introduction to *Hadewijch: The Complete Works*, translation and introduction by Mother Columba Hart, O.S.B. (Paulist Press, 1980), pp. 1–42.

2. Unless noted otherwise, the translation I give of Hadewijch's text is Hart's. Where I offer my own translation, it is based on Van Mierlo's critical edition and at times helped by Hart's translation and modern Dutch translations of parts of Hadewijch's work by F. Van Bladel, S.J., and B. Spaapen, S.J. (1954), Paul Mommaers (1979), H. W. J. Vekeman (1980), as well as by the French translation by Jean-Baptiste Porion, O. Cart. (1972). I have at times consulted the critical edition of "Ms. 941 of de Bibliotheek der Rijksuniversitet te Gent" by H. W. J. Vekeman: *Het Visioenenboek van Hadewijch* (Nijmegen: Dekker& Van De Vegt, 1980).

3. Within a "vision" or letter of Hadewijch, I indicate lines cited by giving the number of the paragraph (Hart's paragraphing) in which the lines are contained. Hart's paragraph number corresponds to Van Mierlo's number for the first line in the paragraph.

4. E.g., Thomas Aquinas, *Summa Theologiae* I–II, 3–5. In life after death, the "beatitude of the blest" (*beatitudo beatorum*) is their "complete possession or fruition of supreme good" (*perfecta possessio vel fruitio summi boni*) (5, 2, c.). Conditions of earthly human life render impossible that a human being have here "true and complete human beatitude" (*perfecta et vera beatitudo*) for it consists of the seeing and consequent loving and enjoying of God's essence (5, 3).

147

Still some in their earthly life are rightly called "blest" (*beati*) because, though they do not see the divine essence, they have a share of that beatitude by a certain "fruition" (*fruitione*) of God, supreme good (5, 3, ad 1 and 2, 3, 5). Moreover, God, by a miracle, can grant that individuals see his essence even in this life (I, 12, 11, c. and ad 2; II–II, 175, 3).

5. I follow Hart in capitalizing "Love" when it refers to divine love, and not capitalizing it when it is a creature's love, even when Hadewijch has evidently in mind that this human love is but a participation in Divine Love. The word being translated as "Love" or "love" is usually *minne*, a noun of feminine gender and the appropriate pronoun for it, as Hart preserves in English and as I follow, is feminine: "she" or "her." Nevertheless, for Hadewijch, *minne* is God's nature and being; it is, therefore, neither male nor female.

6. Translations of Bernard's text are, unless otherwise indicated, mine. I follow the critical edition, *S. Bernardi Opera* (SBO), in which the sermons of Bernard on the Song of Songs are in volumes I (Sermones I–XXXV) and II (Sermones XXXVI–LXXXVI). Various texts and small works are in Vol. III. I have consulted and profited from the English translation of the Cistercian Fathers Series, but it is too free for the present study.

7. In comparing Bernard of Clairvaux with Hadewijch, I have read widely in but not combed all of Bernard's voluminous *opera*. I have been aided in finding pertinent pages of Bernard, particularly on mystical experience, by work of Robert Linhardt (1923), Étienne Gilson (1940), Michael Casey (1988), John R. Sommerfeldt (1991), and Jean Leclercq (1979, 1982, 1987). On the other hand, I put to Bernard's text questions that these scholars did not. My goal in the present essay is not to render a definitive interpretation of Bernard, no more than of Hadewijch. My goal is to launch scholarly inquiry and discussion in a promising direction.

8. Hadewijch affirms that in her supreme union with the Divine she "is" the Divine in, e.g., Vision 1, 138ff.; Vision 3, 1ff.; Vision 7, 1ff.; Vision 14, 145ff.; and Letter 12, 1ff. Similarly, see Bernard, e.g., *De Diligendo Deo* X, 27–28 (SBO III, 142–43).

9. Étienne Gilson (1940) defends Bernard lengthily and convincingly of the pantheism of which he has been accused. For evidence of Gilson's thesis, that for Bernard the soul in its supreme union with God is still not God, see, e.g., SBO II, LXXI, 7–10; LXXXIII, 5; *De Diligendo Deo*, X, 27–28: SBO III, 142–43.

## Chapter 2

1. Cf. Leclercq (1987, 49) on Bernard's thought: "The soul must receive a special charism, that gift of the Holy Spirit which St. Paul refers to as 'the discerning of spirits,' in order to recognize the voice of the Word....For the

thoughts inspired by the Bridegroom are similar to our own, since they call into play the same powers of the mind. Their source, however, is different: They come from God, not from us...[God] reveals himself through the intimacy of love. He is received into the innermost heart, which he fills with his loving presence. He does not appear, he enters the soul. He touches and excites the heart, communicating his love without saying much, for his effects speak of his presence."

## Chapter 3

1. Italicized in the quotation are my changes to Hart's translation: "contents" and "contentedness" for Hart's "suffices to" and "self-sufficiency." The original is *ghenoech* and *ghenoechte*, which Hart generally, and earlier in this passage, translates "content." One could translate it equally well as "satisfies" and "satisfaction."

2. "...cum amat Deus, non aliud vult, quam amari: quippe non ad aliud amat, nisi ut ametur, sciens ipso amore beatos, qui se amaverint." SBO II, LXXXIII, 4 (301, ll. 10–11).

3. SBO II, LXXI, 5. Cf. Roger De Ganck (1991a, 313–17). DeGanck quotes also *Poem in Couplets* 16, 31–32. In the Eucharistic union "The heart of each devours the other's heart / One soul thrusts itself impetuously into the other..." (De Ganck's translation, more literal than Hart's).

## Chapter 4

1. E.g. Martin Luther's preface to the New Testament and preface to the Epistle of St. Paul to the Romans, in *The Freedom of a Christian* (1961, 14ff., 19ff., 42ff.).

## Chapter 5

1. Hart (1980) 366, n.30. Hart follows Porion (1972, 9, 40), who dates Beatrice as living a little later than Hadewijch. Cf. Roger De Ganck, (1991b, 149–50).

## Chapter 6

1. Letter 7. *Minne* is a favored word of courtly love. See Hart (1980, 8).

2. Letter 2, 106ff. For Hadewijch *trouw* and *ontrouw* can mean "faithful-

ness" and "unfaithfulness" (not "faith" and "unfaith"). Elsewhere I argue that Hart favors too much the second pair of meanings, but in this context, I agree with her translation. In *Stanzaic Poem 25*, reason impedes the soul's complete union with the Beloved in that reason, "judges the loved soul and the Beloved ever unequal." But reason is thus "Love's surgeonness" wounding the soul to heal its shortcomings and eventually saying, "Behold! Take possession of the highest glory!"

## Chapter 8

1. See in the bibliography my recent attempts to mark the relevance and range of bodily affective knowing in reproductive choices, in college education, in social concern, and in sexual and other interpersonal relations. So too, my earlier, more general arguments for an ethical epistemology that universally endorses such knowing. In a recent essay, I build on the study of this second half of the book. The essay is "Ethics and Another Knowing of Good and Evil," *Annual. Society of Christian Ethics 1991*, pp. 237–48. The second half of the book revises and expands my "A Medieval Lesson on Bodily Knowing: Women's Experience and Men's Thought," *Journal of the American Academy of Religion* 57/2 (Summer 1989): 341–72.

2. This quotation is in Bernard McGinn, *The Language of Love in Christian and Jewish Mysticism*, unpublished paper, 1986, p. 3. He translates Rupert of Deutz, *Commentary on Matthew* (*The Victory of the Word of God*, Book 12), in J.-P. Migne, *Patrologia Latina*, vol. 168, col. 1601.

## Appendix

1. For the framework of Thomas's special ethics, see the prologue and table of contents of his *Summa Theologiae*, II–II. Bailey (1959, 158–59) finds in this part of the *Summa* the fullest medieval treatment of questions about sexual morality.

2. ST II–II, 141, 3 and 4; 142, 1 and 4; 151, 2 and 3; 152, 1 and 3; 153, 1, 3, 4, and 5; 154, 1. At times, Thomas expresses the object of these virtues and vices not as the pleasure but as the "appetite" (*concupiscentia* or *appetitus*). These are convertible terms, since Thomas defines pleasure simply as the resting of appetite in the possession of its desired good (e.g., *Summa Theologiae*, I–II, 2, 6; 4, 1 and 2; 31, 1 and 2). Not all virtues and vices of Thomas's special ethics are specified by particular pleasures and corresponding desires. For example, justice and injustice concern directly not any pleasure or passion but certain human actions (ST II–II, 58, 9).

3. That Thomas's view of sexual pleasure is the coherent, limited one I claim to find in his text is further supported by comparison with a source he

draws on repeatedly for his ethics: Aristotle. Essentially the same consistent, narrow understanding of sexual pleasure appears in the *Nicomachean Ethics*.

This should also give matter for reflection to those who explain Thomas's view of sex by motifs from other sources: Manichaean dualism or the mysticism, eschatology, and otherworldliness of Augustine and other Christian and pagan Platonists, or Christian dogmas concerning divine law, the purpose of sex and marriage, the fallen nature of man, the evil of sin, the supernatural end of man, etc. All these are lacking in Aristotle. This suggests that the understanding of sexual pleasure common to Thomas and Aristotle may determine Thomas's sexual ethics more than the other motifs do.

The limits of space of the present essay prevent me from doing more than referring to the more evident parallels in the *Nicomachean Ethics* (hereafter: NE) as I proceed with analysis of the Thomistic text.

4. The broad lines of the argument and conclusions are also Aristotle's. Aristotle does not restrict morally permissible sex to marriage. His idea of spiritual activity does not include prayer, worship, or the Eucharist. But, as Thomas notes, his argumentation is basically the same: even in permissible sexual activity, the pleasure so absorbs the mind as to conflict with the higher activity of the soul. Consequently, the virtuous person limits the exercise of sexual activity. Cf. NE III, 12, 19[b] 6–15; VII, 12, 53[a] 28–37.

5. SCG, III, 125. Unless otherwise indicated, all translations are mine. Cf. ST II–II, 154, 9. Thomas cites NE VI, 40[b] 10–20.

6. The same silence of Thomas can be observed in *De Perfectione Vitae Spiritualis*, 8, where the affection spouses may have for each other is treated solely as an obstacle to perfection in the spiritual life.

7. DV, 25, 5, ad 7; *Super Primam Epistolam S. Pauli ad Corinthios Expositio*, 7, 1; Suppl. 41, 4 and 49, 6; ST II–II, 154, 2, ad 6; *De Malo*, 15, 2, c.; *De Duobus Praeceptis Caritatis et Decem Legis Praeceptis*, "De sexto praecepto legis." Cf. ST II–II, 154, 1. 4; 8, ad 2.

8. DV, 25, 6 and 7; ST I–II, 17, 9; 34, 1; 85, 5; *Super Primam Epistolam S. Pauli ad Corinthios Expositio*, 7, 1; *In Decem Libros Ethicorum Aristotelis ad Nieomachum Expositio* (hereafter: In Eth. Nic.), VII, 6, 1389. Cf. Fuchs 1949, 19, 23, 53–56. *Concupiscentia* is the term Thomas uses for the general human appetite for any sense pleasure, but he sees the effect of the Fall verified preeminently in this appetite inasmuch as it is the appetite for sexual pleasure. See also DV 25, 2; ST I–II, 30. That Thomas is the first medieval theologian to consider concupiscence's independence of reason to be natural may be because he is one of the first medieval theologians to read the NE, e.g., X, 9, 79[b] 3–30 or VII, 14, 1 54[b] 20–31.

9. The vice is a subspecies of the vice of "insensibility." In Eth. Nic., 8, 342; ST III, 21, 630-31; ST II–II, 142, 1, c. and ad 1; 152, 2, ad 2; 153, 3, ad 3; *De Malo*, 15, 1, ad 9. Cf. In II Eth. Nic., 2, 262; SCG III, 136. Cf. NE III, 11, 19[a] 1–20; 12, 19[b] 1–20; VII, 7, 50[a] 17–23.

10. ST I–II, 56, 4, c. and ad 4; 58, 3, ad 2; ST II–II, 155, 4; 156, 1, ad 1; DV, 14, 4, c.; *De Virtutibus in Communi* (in *Quaestiones Disputatae*, II), 4; *Super Primam Epistolam S. Pauli ad Corinthios Expositio*, 7, 1. Since, as was noted above, the human sex appetite after the Fall has a certain irremovable independence of reason, the virtuous ordering by reason is not complete domination. Reason rules the natural appetites, not as slaves, but as free citizens. Cf. NE VII, 9.

11. ST I–II, 24, 3, c. The "sense appetite" (*appetitus sensitivus*) differs from the "rational appetite" (*appetitus intellectivus*) or "will" (*voluntas*) in that the former is moved by what the senses perceive, the latter, by the intellect. See ST I, 80.

12. SCG IV, 83; Suppl., 49, 6; 65, 3, c; cf. CT, 1, 156. On Thomas's restriction of the subjective purposes permissible for conjugal intercourse, see the excellent, complementary accounts of Noonan (1965, 241–57, 284–95) and Fuchs (1949, 208–27).

13. SCG IV, 83. On the basis of this ethical principle, Thomas concludes that it would be wrong to act virtuously on earth in order to have sexual pleasure in heaven. The virtuous Christians would, in this hypothesis, have these pleasures somehow as their intention (*in intentione eorum...aliqualiter delectationes praedictae*) and end in view (*finem*). What is wrong is not that they would be intending sexual pleasure alone, for presumably the virtuous Christians would be also intending procreation in their intercourse. What is wrong is that they would be intending sexual pleasure at all "for its own sake" (*propter seipsam*). Cf. Fuchs 1949, 212, 226. For Aristotle, see note 20 below.

14. SCG, IV, 86; Suppl., 82, 3, ad 4; ST I–II, 4, 6; CT, 1, 167; *De Virtutibus in Communi*, 4, 8 and ad 8.

15. SCG IV, 83. In the state of beatitude, there is no longer place for the two purposes of conjugal intercourse permissible in a person's terrestial life: procreation and assistance to the spouse in avoiding sin. Thomas premises, therefore, that if the blest would have sexual intercourse, it could only be for the pleasure of it.

16. ST II–II, 142, 1; 150, 1, ad 1; 152, 2, ad 2; 153, 3, ad 3; In II Eth. Nic. 8, 342; ST III, 21, 630–31. Cf. NE III, 11, 19[a] 5–8; VII, 9, 51[b] 22–52[a] 7; VII, 14, 54[a] 17–19.

17. ST II–II, 153, 2, ad 2; Suppl., 41, 3, ad 6; 49, 4, ad 3; cf. *Serlptum Super Llbros Sententiarum*. II (hereafter: In Sent.) dist. 20, 1, 2, ad 2; SCG III, 126; ST I–II, 34, 1, ad 1.

18. That Thomas does not merely deny that the sexual pleasure of the marital act is evil but even affirms that it is something good is maintained by competent modern commentators (Fuchs 1949, 219, adducing Suppl., 49, 6; Pieper 1965, 27, adducing ST II–II, 153, 2; Noonan 1965, 293, adducing In Sent. IV, 31, 2, 1, ad 3.) One could cite equally well STI–II, 34, 1, c. It is worth noting,

however, as these commentators do not, that in the cited texts, Thomas does not make the affirmation in so many words. That he makes the affirmation implicitly is a reasonable interpretation, though one must postulate in each text at least one tacit premise, e.g., that no human act is morally indifferent. The remainder of the present article should suggest that Thomas's concept of the goodness of conjugal sexual pleasure was so restricted and refined that he may not have felt inclined to affirm it explicitly as such.

19. Suppl., 49, 1, ad 1; 65, 3, c.; ST I–II, 4, 2, ad 2; SCG III, 26; In II Sent., dist. 38, 1, 2, ad 6.

20. This problem of interpretation posed by Thomas's text leads both Fuchs (1949, 226–27) and Noonan (1965, 294–95) to postulate two conflicting currents of Thomas's thought. For both Fuchs and Noonan, Thomas's statement that God intends sexual pleasure to be an inducement contradicts his position that to act for sexual pleasure in marriage is evil. Noonan explains, "But Thomas's statement on inducement was a departure from Aristotelian principle, according to which pleasure itself was always attendant upon some act: one acted for the act itself, the pleasure followed."

But Aristotle himself expresses doubt about this principle (NE X, 4, 75[a] 15–22) and even affirms the contrary: ..."but themselves (for if nothing resulted from them we should still choose each of them)" (NE I, 7, 97[b] 1–5). He raises the same issue as Thomas when he distinguishes: "For not everyone who does anything for the sake of pleasure is either self-indulgent or bad or incontinent, but he who does it for a disgraceful pleasure" (NE VII, 9, 51[b] 18–23).

There are pleasurable things that are "worthy to be chosen for themselves" (NE VII, 4, 47[b] 23–30; 48[a] 22–48[b] 5; VII, 7, 50[a] 16–21; VII, 9, 51[b] 18–23; X, 3, 74[a] 8–12; 1, 7, 97[b] 1–5). But the sexually pleasurable is not such; it is not rightly chosen for itself (NE VII, 4, 47[b] 23–30; VII, 7, 50[a] 16–21; VII, 9, 51[b] 18–23; X, 3, 74[a] 8–11). The person of virtue shuns this kind of pleasure while seeking the higher kind (NE. VII, 11, 53[a] 27–37).

And yet not only should the person of virtue enjoy his or her sexual pleasure (cf. note 16), but he or she naturally will and should desire it (NE III, 11, 18[b] 8–13; 19[a] 15–21; 12, 19[b] 11–20; VII, 14, 54[a] 15–18). Aristotelian principles pose the same dilemma of interpretation as the Thomistic. Sexual pleasure should be enjoyed and desired; it is positively good. Sexual pleasure must not be chosen for itself.

21. ST I–II, 4, 2, ad 2; 3, 4, ad 4; 23, 1; ST I, 81, 2. Cf. ST I–II, 31, 6, ad 3; In VII Eth. Nic., 6, 1389. Cf. NE III, 12, 19[b] 5–20; VII, 3, 47a 31–47[b] 5, 15–18.

22. ST I–II, 2, 5; 4, 2, ad 2; 31, 6; 34, 2, ad 1; 34, 4; ST II–II, 141, 1, 3, 4, 5 and 6; 153, 2 and 3; SCG, IV, 83; DV, 25, 5, ad 7. See also the texts cited in note 19 above. *Pace* Fuchs and Noonan, there is no "contradiction...between the statement that God intends sexual pleasure to be an inducement and the statement that to act for sexual pleasure in marriage is evil" (Noonan, 1965, 294). Sexual pleasure can be understood to properly induce the sense appetite of the

spouses and yet not be suitable to be the spouses' rational purpose. It is no contradiction, but it presses the question of interpretation: What, in the nature of sexual pleasure, the good and proper object of a natural sense appetite, bars it from being a goal for human choice? The question is as pertinent to the Aristotelian text as to the Thomistic.

Under the Pauline category of rendering the spouse what is his or her due (1 Cor. 7:3–5), Thomas acknowledges one other legitimate purpose for conjugal relations: to aid the spouse to avoid sin. He refuses to accept as legitimate and sinless the intent of helping oneself avoid sin, of preserving one's health, and of getting pleasure. Cf. Noonan 1965, 242, 248–49, 284–92, and Fuchs 1949, 138, 200–205, 221–27.

23. ST I–II, 31, 5, c. and 6, c.; *In Metaphysicam Aristotelis Commentaria* I, 1, 5. Cf. ST I–II, 30, 3; 31, 3; 35, 2, ad 3; ST II–II, 180, 7; SCG III, 33; NE III, 10, 18[a] 1-20. To take pleasure in something precisely as known is, by definition, to take pleasure in its beauty. Consequently, humans alone take pleasure in the beauty of what they perceive with their senses (ST I, 91, 3, ad 3; ST I–II, 27, 1, ad 3; cf. CT, 1, 165; ST II–II, 141, 2, ad 3; Kovach 1961, 232–67.) Cf. NE III, 10, 18[a] 1–12, 17–19, 26–30; X, 3, 74[a] 4–8; 74[a] 15–75[a] 2; X, 5, 75[b] 24–76[a] 3. In this line of thought, Thomas often echoes *Metaphysics*, 1, 980[a] 22–28.

24. Inasmuch as a particular good is a *finis* for humans and that "for the sake of" (*propter*) which they act, so, too, one can say, is the pleasure they take in acquiring and possessing the good (ST I–II, 2, 6; 25, 2; 34, 3; In II Sent. dist. 38, 1, 2. c. and ad 6; In VII Eth. Nic., 6, 1389, and X, 6). Cf. NE VII, 12, 52[b] 33–53[a] 15, 30–35; 14, 54[b] 15–20; X, 3, 73[b] 12–18; 74[a] 4–8; X, 4, 74[b] 15–75[a] 2, 10–22; X, 5, 75[b] 24–76[a] 3. Thus, the pleasure of sense perceptions is, in general (though not in the case of sex) one of those pleasures to be "chosen for themselves" (NE VII, 4, 47[b] 23–30; VII, 7, 50[a] 16–21; VII, 9, 51[b] 18–23; X, 3, 74[a] 4–11.

25. SCG IV, 83; cf. Suppl. 49, 6, "Sed Contra"; ST II–II, 148, 2, c.; *De Malo*, 14 (e.g., I and ad 1); 15, 2, ad 6. Cf. NE III, 10; NE VII, 4, 47[b] 23-30; VII, 7, 50[a] 17–21; VII, 9, 51[b] 20–23.

26. ST I–II, 31, 6, c; 60, 5, c.; ST II–II, 141 ff., passim. These pleasures of eating and drinking which Thomas categorizes as pleasures of touch are not pleasures of taste, but the pleasures of satisfying hunger and thirst where taste is not important. The person excessively devoted to these pleasures is not the gourmet but the gourmand or glutton. Cf. NE III, 10.

27. ST II–II, 141, 2 and 4; ST I–II, 31, 6; 60, 5. It does not, however, pertain to temperance to govern pleasures of touch of a higher nature and proper to human beings alone, e.g., the pleasures of the gymnasium (In III Eth. Nic., 20, 617). Similarly, although individuals will not have in the beatific life the pleasures of touch of sex, food, and drink, they will nevertheless have exquisite pleasures of touch (Suppl. 83, 6. ad 3). Indeed, they may have there the pleasure of taste, though not that of satisfying their hunger (Suppl. 82, 4, ad 2). Cf. NE III, 10–12; VII, 4 47[b] 23–30; VII, 12, 52[b] 33–53[a] 15.

28. ST I–II, 34, 1, c. The reader will hopefully pardon the solecism and see the utility in this context of rendering *bonum* and *bona* literally, i.e., as the nouns "good" and "goods." Pleasure, as was seen (note 2) is nothing but the resting of an appetite in a given good. It is true that the goodness of the operation with which the pleasure is connected can also determine the moral goodness of the pleasure (ST I–II, 31, 1, c). But it does so either as being itself a good in which this particular pleasure is taken (cf. ST I–II, 32, 1) or as effecting some further good result. In this latter regard, the moral goodness of a sense pleasure may come from the fact that its corresponding sense activity makes possible intellectual knowledge or the procreation of a new member of the species. However, this latter way of determining the moral goodness of a pleasure does not concern us, since we are comparing only the *intrinsic* moral goodness of sense pleasures, i.e., the moral goodness of the pleasures inasmuch as they would be sought for their own sake. In NE X, 4 and 5, the goodness of different pleasures is determined by the different activities they complete. Aristotle here treats only of the pleasures of sense perception and thought and evaluates these activities according to the condition of the organ and the excellence of the objects known.

29. Thomas evaluates all human's higher activities according to this norm. A person's ultimate end, supreme good, and most perfect activity must be "the best activity" of which he or she is capable (ST I–II, 3, 5). It must, therefore, be the best knowledge of "the best object," God (ST I–II, 3, 4–8). The contemplative life is superior to the active life because the former, unlike the latter, consists in intellectual knowledge (In III Sent., dist. 35, 1, 4; ST II–II, 152, 2; 182, 1). Intellect is superior to sense because it knows "more perfectly" (ST I–II, 2, 6; 31, 5). Cf. NE X, 4, 74$^b$ 15–75$^a$ 2; 75$^b$ 36–76$^a$ 3. Closer Aristotelian parallels can be found outside NE, e.g., *Metaphysics* I, 982$^a$ 3–7.

30. ST I–II, 35, 2 and 7. Thomas draws a conclusion only about the internal sense of "imagination." But he reasons from a universal principle bearing on all internal senses. Cf. ST I, 78, 4; DV, 25, 2, c.

31. In I Metaphys. Aris., 1, 5–9; ST I, 91, 3, ad 3; ST I–II, 31, 6; *De Anima* (in *Quaestiones Disputatae*. II), 13; *In Aristotelis II Librum De Anima Commentarium*, 14, 417. Cf. ST I–II, 27, 1, ad 3; ST II–II, 180, 7, c.; Suppl. 91, 4, c. and ad 1; SCG, III, 53. Cf. NE X, 5, 75$^b$ 36–76$^a$ 3.

32. ST I–II, 27, 1, ad 3; ST I, 5, 4, ad 1. Cf. Kovachs 1961, 232–56 and NE X, 3, 74$^a$ 4–8; X, 4, 74$^b$ 15–20.

33. In NE the pleasure-giving activity in the case of sex, food, and drink is not presented as a knowing activity at all but as the replenishing activity of so much of our state and nature as has remained unimpaired (NE VII, 12, 52$^b$ 33–53$^a$ 18; VII, 14, 54$^b$ 15–20; cf. NE VII, 14, 54$^a$ 35–54$^b$ 2; X, 3, 73$^b$ 7–12). And yet the pleasures of sex, food, and drink are identified by Aristotle as pleasures of the sense of touch (see note 26 above).

34. E.g., the memory and "cogitative" faculty of humans (ST I, 78, 4, ad 5). I use "reason" throughout this essay as Thomas often does, i.e., not distinguishing it from "intellect" but meaning by the term all of a person's spiritual knowing powers (e.g., ST I–II, 58, 2 and 3). In other places, Thomas distinguishes between the lower spiritual knowing capacity of human nature, "reason," and the higher "intellect" or "intelligence" (e.g., ST II–II, 180, 4, ad 3). But he makes clear that these are really just two different aspects of the same *potentia* (e.g., DV, 15, 1; ST I, 79, 8; ST I–II, 31, 7).

35. ST II–II, 141, 7, 1 and ad 1; DV, 15, 1, c.; 25, 2, c.; 6, c. and ad 2. It is the pleasures and appetites of sex, food, and drink that Thomas characterizes as furthest from reason, but the same would have to follow for any sense knowledge essential to the pleasures. Similarly, touch is the most "material" (*maxime materialis*) of all the senses, just as ocular vision is the most spiritual and immaterial (Suppl., 82, 4, ad 1; 91, 4, ad 1; SCG III, 53; *In Aristotelis II Librum De Anima Commentarium* 7, 14, 417).

36. ST II–II, 156, 4, c.; ST I–II, 73, 5, ad 3; *De Malo*, 12, 4, c. and ad 5; In VII Eth. Nic., 6, 1389; cf. DV, 25, 2; ST I–II, 17, 9, ad 3; 30, 3; 31, 7; 46, 5.

37. SCG III, 27, 125, and 137, ad 5; In II Sent., 20, 1, 2, ad 1; ST II–II, 141, 2, ad 2; 153, 1 and 2. Cf. *De Malo* 14, 1, c.; Suppl. 49, 6, ad 4. Correspondingly, it is impossible for a person to have any rational thought during intercourse, so absorbing is the sense pleasure: ST I–II, 34, 1, ad 1; 37, 1, ad 2; ST II–II, 53, 6; 55, 8, ad 1; 153, 2, 2 and ad 2; Suppl., 41, 3, 6 and ad 6; In VII Eth. Nic., 11, 1477. See texts cited in notes 5, 6, 7, 9. Cf. NE VII, 52[b] 17–19, a favorite text of Thomas.

38. SCG, III, 63; IV, 83; ST I–II, 2, 6, "Sed Contra"; 60, S; In Matt. 22 (19: 549[b]). Cf. SCG, III, 27; ST I–II, 3, 5, c.; 17, 9, ad 3; 24, 1, ad 1; 31, 4, ad 3; 31, 5, c.; 32, 7, ad 2; 34, 1, ad 2; ST II–II, 141, 7, ad 1; 182, 1, c.; Suppl., 49, 1, ad 1; In VIII Eth. Nic., 12, 1723. Intemperance is, of all the vices, the most base because it loves as the greatest good a kind of pleasure that animals love, too. The sense of touch giving these pleasures is the most common of the senses; all animals have it. These pleasures are to be distinguished from the pleasures of the gymnasium which are "proper to man and rationally acquired." (In III Eth. Nic., 20, 616–17; ST I–II, 73, 5, ad 3; ST II–II, 142, 4, c.; 179, 2, ad 1 and 3.) Temperance is a particularly beautiful and worthy virtue because it moderates these pleasures "common to us and brutes" (ST II–II, 141, 8, 1 and ad 1; cf. ST II–II, 141, 1, ad 1; 2, ad 3). Since in the act of intercourse these pleasures completely dominate a person, he or she becomes a brute animal at this moment (*bestialis efficitus*; I, 98, 2, ad 3); cf. *Catena Aurea*, In Matt. XXII. Cf. NE III, 10, 18[a] 23–26, 18[b] 3–4; I, 5, 95[b] 19–21; NE VI, 12, 44[b] 8–10; NE VII, 12, 53[a] 28–35; NE X, 9, 80[a] 10–12.

39. ST I–II, 32, 7, ad 2; 34, 1, ad 2; ST II–II 42, 2; 151, 2, ad 2. Cf. NE III, 12, 19[a] 33–19[b] 15; VI, 13, 44[b] 8–10; VII, 12, 53[a] 28–35; X, 6, 11, 76[b] 15–25, 32–34; X, 9, 79[b] 30–35; X, 1, 72[a] 19–22.

40. ST I–II, 30, 5; 46, 2 and 3.

41. ST II–II, 42, 4, c.; SCG, III, 124; In III Eth. Nic., 20, 61. Cf. NE III, 10, 18ᵃ 25–26; 18ᵇ 3–5; III, 11, 18ᵇ 20–21; I, 5, 95ᵇ 19–21; X, 6, 77ᵃ 5–10.

42. ST II–II, 152, 2, ad 2 (*agricola*); In II Eth. Nic., 2, 262 (*homines agrestes*). Cf. NE II, 2, 4ᵃ 23–26.

43. ST I–II, 2, 6, ad 2; 30, 3, c.; 31, 5, ad 1; 32, 7, ad 2. Cf. NE X, 9, 79ᵇ 10ff; VII, 13, 53ᵇ 32–36, X, 1, 72ᵃ 23–26; I, 8, 99ᵃ 12–15.

44. In X Eth. Nic., 6, 2023–29; ST I–II, 21, 7, ad 3; CT I, 165; *In III De Anima.* 1, 2, 598; Suppl. 82, 4, ad 3; In III Eth. Nic., 20, 616–17.

45. E.g., ST I–II, 27, 2, ad 2; 31, 6, ad 2; ST I, 79, 4, 8, 9; 91, 3, ad 3; In Aris. Metaphys., 1, 5–9. Cf. NE X, 5; VII, 14.

46. Cf. Milhaven 1976. I revise and expand the phenomenological analyses of Chirpaz 1969 and 1970.

Aristotle
>    1894    *Ethica Nicomachea.* London: Oxford University Press.
>    1925    *Nicomachean Ethics.* Trans. W. D. Ross. London: Oxford University Press

Augustine
>    1962    *Œuvres de Saint Augustin* 13, *Les Confessions,* 2 vols. Bruges: Desclée de Brouwer.

Aumann, Jordan
>    1967    "Contemplation." In *New Catholic Encyclopedia.* Boston: McGraw-Hill.

Bailey, D. S.
>    1959    *Sexual Relation in Christian Thought.* New York: Harper and Brothers.

Bernard of Clairvaux
>    1957    Sermones super Cantica Canticorum. In *S. Bernardi Opera,* vol. 1. Rome: Editiones Cistercienses.
>    1958    Sermones super Cantica Canticorum. In *S. Bernardi Opera,* vol. 2. Rome: Editiones Cistercienses.
>    1963    *Tractatus et Opuscula.* In *S. Bernardi Opera,* vol. 3. Rome: Editiones Cistercienses.
>    1971    *Cistercian Fathers Series 4. On the Song of Songs I.* Trans. Kilian Walsh. Shannon, Ireland: Cistercian Publications.
>    1979    *Cistercian Fathers Series 31. On the Song of Songs III.* Trans. Kilian Walsh and Irene M. Edmonds. Kalamazoo, Mich.: Cistercian Publications.
>    1980    *Cistercian Fathers Series 40. On the Song of Songs IV.* Trans. Irene M. Edmonds. Kalamazoo, Mich.: Cisterican Publications.

Browning, Don S.
>    1983    *Religious Ethics and Pastoral Care.* Philadelphia: Fortress.

Butler, Cuthbert
1927    *Western Mysticism: The Teachings of SS. Augustine, Gregory, and Bernard on Contemplation and the Contemplative Life*, 2d ed. London: Constable.

Bynum, Caroline Walker
1984    "Women Mystics and Eucharistic Devotion in the Thirteenth Century," *Women's Studies* 11: 179–214.
1985a   "The Veneration of the Eucharist Among Women in the Middle Ages," paper presented at the College of the Holy Cross Symposium, *The World Becomes Flesh: Radical Physicality in Religious Sculpture of the Later Middle Ages.*
1985b   "Fast, Feast, and Flesh; The Religious Significance of Food to Medieval Women," *Representations* 11:1–25.
1986    "The Body of Christ in the Later Middle Ages: A Reply to Leo Steinberg," *Renaisance Quarterly* 39, no. 3 (Autumn): 399–439.
1987    *Holy Feast and Holy Fast: The Religious Significance of Food to Medieval Women.* Berkeley: University of California Press.

Casey, Michael
1988    *A Thirst for God: Spiritual Desire in Bernard of Clairvaux's Sermons on the Song of Songs.* Kalamazoo, Mich.: Cistercian Publications.

Chenu, Marie Dominique
1964    *Toward Understanding St. Thomas.* Trans. A. M. Landry and D. Hughes. Chicago: H. Regnery.
1968    *Nature, Man and Society in the Twelfth Century: Essays on New Theological Perspectives in the Latin West.* Chicago: University of Chicago Press.

Chirpaz, Francois
1969    "Dimensions de la sexualité." *Études* 330:409–23.
1970    "Sexualité, morale et poétique." *Lumière et vie*, 19:72–88.

Colledge, Eric
1961    *The Medieval Mystics of England.* New York: Scribner.

Cunneen, Sally
1991    *Mother Church: What the Experience of Women Is Teaching Her.* New York: Paulist Press.

Davis, Charles
1976    *Body as Spirit: The Nature of Religious Feeling.* New York: Seabury.

De Ganck, Roger
1991a   *The Life of Beatrice of Nazareth, 1200–1268.* Kalamazoo, Mich.: Cistercian Publications.

1991b     *Beatrice of Nazareth in Her Context.* Kalamazoo, Mich.: Cistercian Publications.

1991c     *Towards Unification with God: Beatrice of Nazareth in Her Context.* Kalamazoo, Mich.: Cistercian Publications.

Denzinger, Henricus, and Adolfus Schonmetzer, eds.
> 1963     *Enchiridion symbolorum definitionum et declarationum de rebus fidei et morum.* Freiburg in Breisgau: Herder.

Dionysius the Areopagite: See Pseudo-Dionysius the Areopagite.

Doherty, Dennis
> 1966     *The Sexual Doctrine of Cardinal Cajetan.* Regensburg, Germany: Pustet.

Dreyer, Elizabeth
> 1989     *Passionate Women: Two Medieval Mystics.* New York: Paulist Press.

Fuchs, Josef
> 1949     *Die Sexualethik des heiligen Thomas von Aquin.* Cologne: Bachem.

Ghellinck, Joseph de
> 1948     *Le mouvement théologique du XII<sup>e</sup> siècle.* 2d ed. Bruges: De Tempel.

Gilligan, Carol
> 1982     *In a Different Voice: Psychological Theory and Women's Development.* Cambridge: Harvard University Press.

Gilson, Étienne
> 1929     *The Philosophy of St. Thomas Aquinas.* 2d rev. ed. Trans. Edward Bulough. St. Louis: Herder.
> 1936     *The Spirit of Medieval Philosophy.* Trans. A. C. H. Downes. New York: Scribner.
> 1940     *The Mystical Theology of Saint Bernard.* Trans. A. C. H. Downes. New York: Sheed and Ward.

Goldenberg, Naomi R.
> 1989     "Archetypal Theory and the Separation of Mind and Body. Reason Enough to Turn to Freud?" In *Weaving the Visions: New Patterns in Feminist Spirituality.* Ed. Judith Plaskow and Carol P. Christ. New York: Harper and Row.

Griffin, Susan
> 1978     *Woman and Nature: The Roaring Inside Her.* New York: Harper and Row.
> 1982     *Made from This Earth.* New York: Harper and Row.

Guerric of Igny
    1833      "In Nativitate Sancti Joannis Baptistae Sermo II." In *Patrolo-
              giae cursus completus, Series Latina*, 185, 167–69. Ed. J. Migne.
              Paris: Migne.

Guindon, André
    1986      *The Sexual Creators: An Ethical Proposal for Concerned Chris-
              tians*. Lanham, Md.: University Press of America.

Hadewijch
    1924      *Hadewijch: Visioenen*. Ed. Jozef van Mierlo. Louvain: Vlaam-
              sch Boekenhalle
    1942      *Hadewijch: Strophische Gedichten*. Ed. Jozef van Mierlo.
              Antwerp: Standaard.
    1947      *Hadewijch: Brieven*. Ed. Jozef van Mierlo. Antwerp: Stan-
              daard.
    1952      *Hadewijch: Mengeldichten*. Ed. Jozef van Mierlo. Antwerp:
              Standaard.
    1954      *Hadewijch: Brieven*. Trans. F. Van Bladel and B. Spaapen. Tielt:
              Lannoo.
    1972      *Hadewijch, Lettres spirituelles: Béatrice de Nazareth, Sept degrés
              d'amour*. Trans. Jean-Baptiste Porion, O. Cart. Geneva:
              Claude Martingay.
    1979      *De visioenen van Hadewijch*. Trans. Paul Mommaers.
              Nijmegen: Gottmer.
    1980a     *Het Visioenenboek van Hadewijch*. Ed. and trans. H. W. J. Veke-
              man. Nijmegen: Dekker and Van De Vegt.
    1980b     *Hadewijch: The Complete Works*. Trans. Mother Columba Hart.
              New York: Paulist Press.

Harrison, Beverly
    1985      *Making the Connections: Essays in Feminist Social Ethics*.
              Boston: Beacon Press.

Hart, Mother Columba, O.S.B.
    1980      *Hadewijch: The Complete Works*. Trans. Mother Columba Hart.
              New York: Paulist Press.

Heyward, Carter
    1989      *Touching Our Strength: The Erotic as Power and the Love of God*.
              San Francisco: Harper and Row.

Hodgson, Phyllis
    1944      *The Cloud of Unknowing and The Book of Privy Counselling*.
              London: Oxford University Press.

Hunt, Mary E.
    1991      *Fierce Tenderness: A Feminist Theology of Friendship*. New York:
              Crossroad.

Johnston, William
    1973    *The Cloud of Unknowing and the Book of Privy Counseling.* New York: Doubleday.

Kass, Leon R.
    1985    "Thinking About the Body." In his *Toward a More Natural Science: Biology and Human Affairs,* 276–98. New York: Free Press.

Keen, Sam
    1970    *To a Dancing God.* New York: Harper and Row.

Kovach, Francis J.
    1961    *Die Aesthetik des Thomas von Aquin.* Berlin: De Gruyter.

Leclercq, Jean
    1960    *The Love of Learning and the Desire for God: A Study of Monastic Culture.* Trans. Catharine Misrahi. New York: Fordham University Press.
    1979    *Monks and Love in Twelfth-Century France: Psycho-Historical Essays.* London: Oxford University Press.
    1982    *Monks on Marriage: A Twelfth-Century View.* New York: Seabury.
    1987    *Bernard of Clairvaux: Selected Works.* New York: Paulist.

Linhardt, Robert
    1923    *Die Mystik des heiligen Bernhard von Clairvaux.* Munich: Natur und Kultur.

Lorde, Audre
    1976    *Coal.* New York: Norton.
    1984    "Uses of the Erotic: The Erotic as Power." In *Sister Outsider: Essays and Speeches.* Freedom, Calif.: Crossing Press.

Luther, Martin
    1961    *Martin Luther: Selections from His Writings.* Ed. John Dillenberger. New York: Doubleday.

Maguire, Daniel C.
    1978    *The Moral Choice.* New York: Doubleday.

Mallory, Marilyn May
    1977    *Christian Mysticism: Transcending Techniques. A Theological Reflection on the Empirical Testing of the Teaching of St. John of the Cross.* Amsterdam: Van Gorcen Assen.

Margaret of Oingt
    1990    *The Writings of Margaret of Oingt.* Ed. and trans. Renate Blumenfeld-Kosinski. Newburyport, Mass.: Focus Information Group.

Merton, Thomas
    1977 (1956)    *Thoughts in Solitude*. New York: Noonday.

Miles, Margaret
    1981    *Fullness of Life: Historical Foundations for a New Asceticism*. Philadelphia: Westminster Press.

Milhaven, John Giles
    1970    *Toward a New Catholic Morality*. Garden City, N.Y.: Doubleday.
    1971    "Objective Moral Evaluation of Consequences," *Theological Studies* 32:407–30.
    1974    "Conjugal Sexual Love and Contemporary Moral Theology," *Theological Studies* 35:692–710.
    1976    "Christian Evaluations of Sexual Pleasure," *Selected Papers, 1976, of the Society of Christian Ethics*, 63–74.
    1977    "Thomas Aquinas on Sexual Pleasure," *Journal of Religious Ethics* 5:157–81.
    1978    "The Voice of Lay Experience in Christian Ethics," *Proceedings of the Thirty-third Annual Convention of the Catholic Theological Society of America*, vol. 33, 35–53.
    1984a    "An Experienced Value of Marital Faithfulness in *Dubin's Lives*," *Journal of Religious Ethics* 12:82–96.
    1984b    "The Role of the Affective in the Moral Life," *Proceedings of the Thirty-Ninth Annual Convention of the Catholic Theological Society of America*, vol. 39, 163–65.
    1985    "Getting Emotional in Class," *George Street Journal* (Brown University) 10, no, 12 (March 5): 7.
    1986a    "How Prolife Beliefs Strengthen a Prochoice Position," *Conscience* 7, no. 3: 17–18.
    1986b    "The Problems with an Anti-Choice Amendment," *Conscience* 7, no. 4:18.
    1987    "Getting Emotional in Class—II," *George Street Journal* 12/11 (February 18): 16.
    1988a    "How the Church Can Learn from Gay and Lesbian Experience." In *The Vatican and Homosexuality*, 216–23. Ed. Jeannine Gramick and Pat Furey. New York: Crossroads.
    1988b    "Becoming Prolife While Staying Prochoice," *Conscience* 9, no. 5: 15–18.
    1989a    "Sleeping LIke Spoons. A Question of Embodiment," *Commonweal* 116, no. 7: 205–7.
    1989b    *Good Anger*. Kansas City, Mo.: Sheed and Ward.
    1989c    "A Medieval Lesson on Bodily Knowing: Women's Experience and Men's Thought," *Journal of the American Academy of Religion* 57, no. 2 (Summer): 341–72.
    1991    "Ethics and Another Knowing of Good and Evil," In *Annual of the Society of Christian Ethics*, 237–48.

Müller, Max
    1958    *Existenzphilosophie im geistigen Leben der Gegenwart*, 2d ed. Heidelberg: Kerle.

Nelson, James B.
    1978    *Embodiment: An Approach to Sexuality and Christian Theology.* Minneapolis: Augsburg.
    1983    *Between Two Gardens: Reflections on Sexuality and Religious Experience.* New York: Pilgrim Press.
    1988    *The Intimate Connection: Male Sexuality, Masculine Spirituality.* Philadelphia: Westminster Press.

Noonan, John T., Jr.
    1965    *Contraception. A History of Its Treatment by the Catholic Theologians and Canonists.* Cambridge: Harvard University Press.

O'Brien, Elmer (ed.)
    1964    *Varieties of Mystic Experience.* New York: New American Library.

Origen
    1966    *Homélies sur le Cantique des Cantiques.* Trans. Olivier Rousseau. Paris: Cerf.

Phipps, William E.
    1974    "The Flight of the Song of Songs," *Journal of the American Academy of Religion* 62:87.

Pieper, Joseph
    1965    *The Four Cardinal Virtues: Prudence, Justice, Fortitude, Temperance.* Trans. Richard and Clara Winston. New York: Harcourt, Brace and World.

Plaskow, Judith
    1991    *Standing Again at Sinai.* New York: Harper Collins.

Plato
    1914    *I. Euthyphro, Apology, Crito, Phaedo, Phaedrus.* Trans. W. R. M. Fowler. Cambridge: Harvard University Press.
    1925    *V. Lysis, Symposium, Gorgias.* Trans. W. R. M. Lamb. Cambridge: Harvard University Press.

Plotinus
    1924ff.    *Ennéades.* Ed. and trans. Émile Bréhier. Paris: Belles Lettres.
    1953    *Plotinus.* Ed. and trans. A. H. Armstrong. London: George Allen and Unwin.

Porion, Jean-Baptiste
    1972    *Hadewijch, Lettres Spirituelles; Béatrice de Nazareth, Sept degrés d'amour.* Trans. Jean-Baptiste Porion and O. Cart. Geneva: Claude Martingay.

Pseudo-Dionysius the Areopagite
  1940    *The Divine Names and The Mystical Theology*. Trans. C. E. Rolt.
          London: S.P.C.K.

Rich, Adrienne
  1976    *Of Woman Born*. New York: Norton.

Richard of Saint-Victor
  1957    *Of the Four Degrees of Passionate Charity* in *Richard of Saint Vic-
          tor: Selected Writings on Contemplation*. Ed. and trans. Clare
          Kirchberger. New York: Harper and Row.

Riehle, Wolfgang
  1981    *The Middle English Mystics*. Trans. Bernard Standring. Lon-
          don: Routledge and Kegan Paul.

Ruether, Rosemary
  1983    *Sexism and God Talk: Toward a Feminist Theology*. Boston: Bea-
          con Press.

Sommerfeldt, John R.
  1991    *The Spiritual Teachings of Bernard of Clairvaux*. Kalamazoo,
          Mich.: Cistercian Publications.

Thielicke, Helmut
  1964    *The Ethics of Sex*. Trans. John W. Doberstein. New York:
          Harper and Row.

Thomas Aquinas
  1689    "Super Primam Epistolam S. Pauli ad Corinthios Expositio."
          In *In Omnes D. Pauli Apostoli Epistolas Commentaria*. Lyons:
          Briasson.
  1927a   "Compendium Theologiae." In *Opuscula Omnia*, II. Paris:
          Lethielleux.
  1927b   "De Duobus Praeceptis Caritatis et Decem Legis Praecep-
          tis." In *Opuscula Omnia* IV. Paris: Lethielleux.
  1927c   "De Perfectione Vitae Spiritualis." In *Opuscula Omnia*, IV.
          Paris: Lethielleux.
  1929    *Scriptum Super Libros Sententiarium*. Paris: Lethielleux.
  1931a   "De Anima." In *Quaestiones Disputatae*, II. Turin: Marietti.
  1931b   "De Malo." In *Quaestiones Disputatae*, II. Turin: Marietti.
  1931c   "De Virtutibus in Communi." In *Quaestiones Disputatae*, II.
          Turin: Marietti.
  1931d   "De Veritate." In *Quaestiones Disputatae*, III and IV. Turin:
          Marietti.
  1934    *Summa Contra Gentiles*. Rome: Vatican.
  1935a   *In Aristotelis Libros de Anima Commentarium*. Turin: Marietti.
  1935b   *In Metaphysicam Aristotelis Commentaria*. Turin: Marietti.
  1948a   *Summa Theologiae*, II–II. Turin: Marietti.

| | |
|---|---|
| 1948b | *Summa Theologiae*, III. Turin: Marietti. |
| 1948c | *Supplementum Summae Theologiae*. Turin: Marietti. |
| 1949 | *In Decem Libros Ethicorum Aristotelis ad Nicomachum Expositio*. Turin: Marietti. |
| 1950a | *Summa Theologiae*, I. Turin: Marietti. |
| 1950b | *Summa Theologiae*, I–II. Turin: Marietti. |

Trask, Haunani-Kay
    1986        *Eros and Power: The Promise of Feminist Theory.* Philadelphia: University of Pennsylvania Press.

Trethowan, Illtyd
    1974        *Mysticism and Theology. An Essay in Christian Metaphysics.* London: Geoffrey Chapman.

Underhill, Evelyn
    1930 (1911)  *Mysticism: A Study in the Nature and Development of Man's Spritual Consciousness*, 12th ed. New York: Dutton.

Van der Marck, William H.
    1967        *Towards a Christian Ethic.* New York: Newman.

Ziegler, Joanna E.
    1985a       *The Word Becomes Flesh: Radical Physicality in Religious Sculpture of the Later Middle Ages* (ex. cat.), Worcester, Mass.: Cantor Art Gallery.
    1985b       "The Virgin or Mary Magdalen? Artistic Choices and Changing Spiritual Attitudes in the Later Middle Ages," paper presented at the symposium, *The Word Becomes Flesh: Radical Physicality in Religious Sculpture of the Later Middle Ages.* College of the Holy Cross.
    1986a       "Women of the Middle Ages. Some Questions Regarding the Beguines and Devotional Art," *Vox Benedictina* 3, 4 (October), 338–57.
    1986b       "The Emergence of a Women's Sensibility in Late Medieval Art in Northern Europe," lecture at Brown University.
    1987        "The *curtis* beguinages in the Southern Low countries and art patronage: interpretation and historiography." *Bulletin de l'Institut Historique Belge de Rome* 57:31–70.
    1988        *The Beguines: Art and the Erotic in Women's Religious Community*, lecture at Union Theological Seminary, New York City.
    1989        "The Medieval Virgin as Object: Art or Anthropology?" *Journal of Historical Reflections/Réflexions Historiques* 16, nos. 2 and 3: 251–64.

1990    "Phenomenal Religion in the Thirteenth Century and Its Image: Elisabeth of Spalbeek and the Passion Cult," with W. Simons. In *Women in the Church* (*Studies in Church History,* 27), 117–26. W. J. Sheils and D. Woods. London: Oxford University Press.

1992    "Reality as Imitation: The Dynamics of Imagery Among the Beguines." In *Maps of Flesh and Light: New Perspectives on Religious Experience of Late Medieval Women,* ed. U. Wiethaus. Syracuse, N.Y.: Syracuse University Press.

1993    *Sculpture of Compassion: The Pietà and the Beguines in the Southern Low Countries.* Belgian Historical Institute of Rome.

Agnes of Montepulciano, 82, 102
Albert the Great, 84
Angela of Foligno, 82
Aquinas, Thomas. *See* Thomas
  Aquinas
Aristotle, 31, 35, 84, 98, 107, 112–113,
  117, 139–140, 142
Augustine, 6–7, 9–11, 13, 25, 38, 84,
  88, 112, 116
Aumann, J., 86

Bailey, D. S., 131
Barth, Karl, 116
Beatrice of Nazareth, ix, 7, 47, 61, 112
Berengarius Turonensis, 91
Bernard of Clairvaux, xii, 6–8, 12–15,
  17–18, 22–26, 33–34, 39–40, 42, 44, 47,
  49, 51, 58, 61, 66–70, 79–80, 84–85, 88,
  103, 106, 108, 112, 114, 118, 139
Blumenfeld-Kosinski, Renate, 82, 103
Boethius, 92
Bonaventure, 7, 40
Bonhoeffer, Dietrich, 43
Browning, Don, 76
Butler, Cuthbert, 84
Bynum, Caroline Walker, 4–6, 77–79,
  81–83, 86, 88, 102, 112, 117

Camus, Albert, 39, 42–43
Cassian, 84
Chenu, Marie Dominique, 4
*Cloud of Unknowing*, author, 66, 85,
  112

Colledge, Eric, 85

Davis, Charles, 86
Denziger, Henricus, 91
Descartes, Réné, 43, 141
Doherty, Dennis, 125
Dreyer, Elizabeth, 67

Elisabeth of Spalbeek, ix

Faral, Edmond, 114
Fra Angelico, 96–97
Francis of Assisi, 79, 118
Fuchs, Josef, 125, 132

Galileo, 141
Gallus, Thomas, 112
Gertrude of Delft, 82, 102
Gertrude of Helfta, 82, 102
Ghellinck, Joseph de, 4
Gilligan, Carol, ix
Gilson, Étienne, 4, 49, 112
Giotto, 96–97
Goldenberg, Naomi, ix, 75
Gregory the Great, 84
Griffin, Susan, ix, 75, 118
Grunewald, 81, 96
Guerric of Igny, 7, 53–57, 61, 67
Guindon, Andre, 76

Harrison, Beverly, ix
Hart, Mother Columba, O.S.B., 4, 11,
  19–21, 29, 47, 50, 57–59, 62

Hegel, G. W. F., 40, 70, 141
Heidegger, Martin, 95
Heyward, Carter, ix, xi, 32, 44, 75
Hildegard of Bingen, 7, 14, 61
Hilton, Walter, 85, 87
Hodgson, Phyllis, 84, 85, 112
Hugh of Saint Victor, 7, 26, 84
Huizinga, Johan, 81
Hunt, Mary, ix, 32, 44, 75

Ida of Louvain, 82, 102

Jacob, 53–54, 88
Jerome, 6, 47
John the Apostle, 18, 24
John of the Cross, 7
John of Ruusbroec, 4
Johnston, William, 84
Julian of Norwich, 7, 87

Kant, Immanuel, xii, 43, 67, 70, 116, 141
Kass, Leon, 76
Keen, Sam, 76
Kempe, Margery, 87–88
Kirchberger, Clare, 66

Langmann, Adelheid, 82
Leclercq, Jean, 4, 22, 88, 114, 115, 139
Lidwina of Schiedam, 82, 102
Lorde, Audre, ix, 44, 75, 112
Luke the Evangelist, 88, 115, 138
Luther, Martin, xii, 42–43, 141

McGinn, Bernard, 79
Maguire, Daniel, 76
Mallory, Marilyn, 77
Marcus Aurelius, 116
Margaret of Cortona, 93
Margaret of Faenza, 82, 102
Margaret of Oingt, ix, 7, 15, 82, 88–89, 99–100, 102–103
Mark the Evangelist, 139
Mary, Mother of Jesus, 4, 58, 116
Mary of Oignies, 93, 112
Matthew the Evangelist, 139

Maximus Confessor, 84
Mechtild of Magdeburg, 7, 15, 87, 93, 139
Merton, Thomas, 84
Michelangelo, 97
Miles, Margaret, 92, 94, 107
Milhaven, J. Giles, 112, 115–116, 145
Mommaers, Paul, 5, 20
Monica, 38
Morgan, Robin, 118
Moses, 55
Muller, Max, 143

Nelson, James, 76
Newton, Isaac, 141
Nietzsche, Friedrich, 43
Noonan, John T., Jr., 125

O'Brien, Elmer, 139
Origen, 22, 84, 106

Paul the Apostle, 86–87, 139
Peter the Apostle, 86–87
Philo, 37
Pieper, Joseph, 130
Plaskow, Judith, ix, 75
Plato, 31, 34, 40, 103–104, 108, 112, 116
Plotinus, 11, 17–18, 22, 84, 103–106, 108
Porphyry, 104
Pseudo-Dionysius, 6–7, 9–14, 22, 25, 38, 40, 69, 84, 106

Rahner, Hugo, 116
Rahner, Karl, 43, 116
Rich, Adrienne, ix, 75, 118
Richard of Saint Victor, 6–7, 12, 26, 58, 61, 66–68, 71
Riehle, Wolfgang, 87–88
Riemenschneider, 81, 96–97
Rodin, Francois, 116
Rousseau, Jean-Jacques, 141
Ruether, Rosemary, ix, 75
Rupert of Deutz, 79

Sappho, 95
Sartre, Jean-Paul, xii, 39, 42–43, 67, 70
Schonmetzer, Adolfus, 91
Socrates, 40, 42, 77, 104
Spaapen, B., 20
Stevens, Emily, 119
Suso, Henry, 79

Teresa of Avila, 7
Thielicke, Helmut, 138
Thomas Aquinas, 7, 20, 31, 34, 39–41,
   44–51, 67–68, 70–72, 84–86, 92–94,
   98, 106–108, 110, 112–113, 116–117,
   123–145

Trask, Haunani-Kay, ix, 75, 118
Trethowan, Illtyd, 86

Underhill, Evelyn, 5, 85

Van Bladel, F., 20
Van der Marck, William H., 125
Van Mierlo, Jozef, 5, 20

William of Saint Thierry, 26

Ziegler, Joanna, 4–6, 77–83, 88, 95, 97,
   101–102, 110, 117